When I ⸻ e picture I had presented in *A Universe from Nothing* by focusing in this volume not on our understanding of the universe on the largest scales, but rather on the largely unheralded (among the public, at least) revolutionary developments over the past half-century in our fundamental understanding of the particles and forces that make up the universe on its smallest scales. In so doing, I wanted to take back another existential question that had been usurped by religion, but which our updated understanding of nature had shed vital new light on—namely, *Why are we here?* This important question is of profound and general interest, and we cannot leave it to the theologians.

In a related vein, by the title, I wanted, by the use of "... So Far," to contrast the remarkable evolving story that science is telling us about the Universe to the static version developed thousands of years ago, which I recently heard aptly described not as "The Greatest Story Ever Told" but rather as "The Goat Herders Guide to the Universe." For, after all, it was a view of the Universe attempted by individuals who didn't even know the Earth orbited the Sun. But I never intended, nor do I still intend, for this book to be a book about religion. First and foremost, I don't mention God because she never really enters in our modern understanding of nature. There is not only no room for supernatural shenanigans; there is no evidence for them.

Nevertheless, some commentators have taken umbrage at my use of section headings—Genesis, Exodus, and Revelations—and my use of bible quotes to begin each chapter. This surprised me. I actually meant no disrespect here. As a work of literature, the bible is full of interesting quotes about many questions of general interest, and I found it poignant that it was easy to find quotes that seemed particularly relevant to the story I was telling. I am not one to deny our cultural heritage, and the bible has certainly played a key role in the development of Western society. And Genesis, Exodus, and Revelations seemed to perfectly describe the three stages of the grand story I had decided to tell. So, to those who are offended by my use of such quotes, I say, *Lighten up*.

But there was something I hadn't anticipated when I worked on this book over a three-year period from 2013–2016. That would be the election of Donald Trump and the associated demise of civil and factual discourse in the national political arena that accompanied his election. I now believe that one of the most important applied morals of the book relates to politics, and not religion.

The story I tell here shows how in less than 400 years, scientists successfully pierced the illusion of reality we experience on human scales, to reveal the characteristics of a universe that at its fundamental scales is dramatically different. In so doing, we can now see that all the characteristics of our universe that make it appear so well-designed for humans are illusions, accidents of nature that could easily have turned out differently. The significance we ascribe to these characteristics is simply based on our cosmic myopia: we assume the Universe we see directly reflects the true underlying nature of reality. Science has instead dragged the scientific community kicking and screaming to a different understanding of nature—one where such things as mass itself, which is responsible for the existence of all the structures we hold dear in the Universe, is a byproduct of a cosmic accident related to the "freezing" of a field throughout space in the early universe.

This remarkable discovery came about by a long and convoluted road of discovery, which I have enjoyed describing in this book. But from the perspective of 2018, what I think is particularly timely about this story is that it demonstrates that if the scientific method can work so effectively to cut through illusions in our view of the Universe, it can also work effectively to cut through illusions that politicians and others attempt to fabricate about the world so they can push through policies based on ideology, and not facts.

We hear often about "facts" and "alternative facts." Part of the problem here is that we somehow give priority to facts, but not to the process by which facts are derived. That process involves the scientific method, and it involves skeptical inquiry, testing our ideas, retesting them, looking at multiple sources and not just one experiment, etc. If we could all apply the same process to our examination of political claims, politicians who regularly lie about the world would have far less success. We need to base public policy on empirical reality, whether or not we like reality. Science, seen as a process, is the way to do that.

The story I have told here demonstrates the scientific process at its best, in my opinion. I hope you enjoy the story, and the awesome realities about nature it reveals, as we have moved beyond the shadows of reality that we evolved to be able to picture, to the deeper aspects of the world that this evolution, as a side benefit, has allowed us to uncover. And let's use the same skills to help us cut through nonsense in the political arena, and in so doing, let the *Greatest Story Ever Told . . . So Far* help the human story to keep getting better every day, and not retreat as a result of myth, superstition, hatred, and ideology.

– *Lawrence M. Krauss*

ALSO BY LAWRENCE M. KRAUSS

A Universe from Nothing

The Fifth Essence

Fear of Physics

The Physics of Star Trek

Beyond Star Trek

Hiding in the Mirror

Quintessence

Atom

Quantum Man

THE GREATEST STORY EVER TOLD ...SO FAR

WHY ARE WE HERE?

Lawrence M. Krauss

**SIMON &
SCHUSTER**

London · New York · Sydney · Toronto · New Delhi

A CBS COMPANY

First published in Great Britain by Simon & Schuster UK Ltd, 2017
This paperback edition published by Simon & Schuster UK Ltd, 2018
A CBS COMPANY

1 3 5 7 9 10 8 6 4 2

Simon & Schuster UK Ltd
1st Floor
222 Gray's Inn Road
London WC1X 8HB

www.simonandschuster.co.uk
www.simonandschuster.com.au
www.simonandschuster.co.in

Simon & Schuster Australia, Sydney
Simon & Schuster India, New Delhi

A CIP catalogue record for this book is available from the British Library

Paperback ISBN: 978-1-4711-3855-3
eBook ISBN: 978-1-4711-3856-0

Interior design by Dana Sloan

Printed and bound by by CPI Group (UK) Ltd, Croydon, CR0 4YY

MIX
Paper from
responsible sources
FSC® C020471

Simon & Schuster UK Ltd are committed to sourcing paper
that is made from wood grown in sustainable forests and support the Forest
Stewardship Council, the leading international forest certification organisation.
Our books displaying the FSC logo are printed on FSC certified paper.

For Nancy

*These are the tears of things,
and the stuff of our mortality
cuts us to the heart.*

—VIRGIL

CONTENTS

Part Three: **Revelation**

THE GREATEST

STORY EVER TOLD

...SO FAR

PROLOGUE

The hardest thing of all to see is what is really there.
—J. A. BAKER, *THE PEREGRINE*

*I*n the beginning there was light.

But more than this, there was gravity.

After that, all hell broke loose. . . .

This is how the story of the greatest intellectual adventure in history might properly be introduced. It is a story of science's quest to uncover the hidden realities underlying the world of our experience, which required marshaling the very pinnacle of human creativity and intellectual bravery on an unparalleled global scale. This process would not have been possible without a willingness to dispense with all kinds of beliefs and preconceptions and dogma, scientific and otherwise. The story is filled with drama and surprise. It spans the full arc of human history, and most remarkably, the current version isn't even the final one—just another working draft.

It's a story that deserves to be shared far more broadly. Already in the first world, parts of this story are helping to slowly replace the myths and superstitions that more ignorant societies found solace in centuries or millennia ago. Nevertheless, thanks to the directors George Stevens and David Lean, the Judeo-Christian Bible is still sometimes referred

1

to as "the greatest story ever told." This characterization is astounding because, even allowing for the frequent sex and violence, and a bit of poetry in the Psalms, the Bible as a piece of literature arguably does not compare well to the equally racy but less violent Greek and Roman epics such as the *Aeneid* or the *Odyssey*—even if the English translation of the Bible has served as a model for many subsequent books. Either way, as a guide for understanding the world, the Bible is pathetically inconsistent and outdated. And one might legitimately argue that as a guide for human behavior large swaths of it border on the obscene.

In science, the very word *sacred* is profane. No ideas, religious or otherwise, get a free pass. For this reason the pinnacle of the human story did not conclude with a prophet's sacrifice two thousand years ago, any more than it did with the death of another prophet six hundred years later. The story of our origins and our future is a tale that keeps on telling. And the story is getting more interesting all the time, not due to revelation, but due to the steady march of scientific discovery.

Contrary to many popular perceptions, this scientific story also encompasses both poetry and a deep spirituality. But this spirituality has the additional virtue of being tied to the real world—and not created in large part to appease our hopes and dreams.

The lessons of our exploration into the unknown, led not by our desires, but by the force of experiment, are humbling. Five hundred years of science have liberated humanity from the shackles of enforced ignorance. By this standard, what cosmic arrogance lies at the heart of the assertion that the universe was created so that we could exist? What myopia lies at the heart of the assumption that the universe of our experience is characteristic of the universe throughout all of time and space?

This anthropocentrism has fallen by the wayside as a result of the story of science. What replaces it? Have we lost something in the process, or as I shall argue, have we gained something even greater?

I once said at a public event that the business of science is to make people uncomfortable. I briefly regretted the remark because I worried that it would scare people away. But being uncomfortable is a virtue, not a hindrance. Everything about our evolutionary history has primed our minds to be comfortable with concepts that helped us survive, such as the natural teleological tendency children have to assume objects exist to serve a goal, and the broader tendency to anthropomorphize, to assign agency to lifeless objects, because clearly it is better to mistake an inert object for a threat than a threat for an inert object.

Evolution didn't prepare our minds to appreciate long or short timescales or short or huge distances that we cannot experience directly. So it is no wonder that some of the remarkable discoveries of the scientific method, such as evolution and quantum mechanics, are nonintuitive at best, and can draw most of us well outside our myopic comfort zone.

This is also what makes the greatest story ever told so worth telling. The best stories challenge us. They cause us to see ourselves differently, to realign our picture of ourselves and our place in the cosmos. This is not only true for the greatest literature, music, and art. It is true of science as well.

In this sense it is unfortunate that replacing ancient beliefs with modern scientific enlightenment is often described as a "loss of faith." How much greater is the story our children will be able to tell than the story we have told? Surely that is the greatest contribution of science to civilization: to ensure that the greatest books are not those of the past, but of the future.

Every epic story has a moral. In ours, we find that letting the cosmos guide our minds through empirical discovery can produce a great richness of spirit that harnesses the best of what humanity has to offer. It can give us hope for the future by allowing us to enter it with our eyes open and with the necessary tools to actively participate in it.

• • •

My previous book, *A Universe from Nothing*, described how the revolutionary discoveries over the past hundred years have changed the way we understand our evolving universe on its largest scales. This change has led science to begin to directly address the question "Why is there something rather than nothing?"—which was formerly religious territory—and rework it into something less solipsistic and operationally more useful.

Like *A Universe from Nothing*, this story also originated in a lecture I presented, in this case at the Smithsonian Institution in Washington, DC, which generated some excitement at the time, and as a result I was once again driven to elaborate upon the ideas I started to develop there. In contrast to *A Universe from Nothing*, in this book I explore the other end of the spectrum of our knowledge and its equally powerful implications for understanding age-old questions. The profound changes over the past hundred years in the way we understand nature at its smallest scales are allowing us to similarly co-opt the equally fundamental question "Why are we here?"

We will find that reality is not what we think it is. Under the surface are "weird," counterintuitive, invisible inner workings that can challenge our preconceptions of what makes sense as much as a universe arising from nothing might.

And like the conclusion I drew in my last book, the ultimate lesson from the story I will tell here is that there is no obvious plan or purpose to the world we find ourselves living in. Our existence was not preordained, but appears to be a curious accident. We teeter on a precarious ledge with the ultimate balance determined by phenomena that lie well beneath the surface of our experience—phenomena that don't rely in any way upon our existence. In this sense, Einstein was wrong: "God" *does* appear to play dice with the universe, or universes. So far we have been lucky. But like playing at the craps table, our luck may not last forever.

* * *

Humanity took a major step toward modernity when it dawned in our ancestors' consciousness that there is more to the universe than meets the eye. This realization was probably not accidental. We appear to be hardwired to need a narrative that transcends and makes sense of our own existence, a need that was probably intimately related to the rise of religious belief in early human societies.

By contrast, the story of the rise of modern science and its divergence from superstition is the tale of how the hidden realities of nature were uncovered by reason and experiment through a process in which seemingly disparate, strange, and sometimes threatening phenomena were ultimately understood to be connected just beneath the visible surface. Ultimately these connections dispelled the goblins and fairies that had earlier spawned among our ancestors.

The discovery of connections between otherwise seemingly disparate phenomena is, more than any other single indicator, the hallmark of progress in science. The many classic examples include Newton's connection of the orbit of the Moon to a falling apple; Galileo's recognition that vastly different observed behaviors for falling objects obscure that they are actually attracted to the earth's surface at the same rate; and Darwin's epic realization that the diversity of life on Earth could arise from a single progenitor by the simple process of natural selection. None of these connections was all that obvious, at first. However, after the relationship comes to light and becomes clear, it prompts an "Aha!" experience of understanding and familiarity. One feels like saying, "I should have thought of that!"

Our modern picture of nature at its most fundamental scale—the Standard Model, as it has become called—contains an embarrassment of riches, connections that are far removed from the realm of everyday experience. So far removed that it is impossible without some grounding to make the leap in one step to visualize them.

Not surprisingly, such a single leap never occurred historically, either. A series of remarkable and unexpected and seemingly unrelated connections emerged to form the coherent picture we now have. The mathematical architecture that has resulted is so ornate that it almost seems arbitrary. "Aha!" is usually the furthest thing from the lips of the noninitiated when they hear about the Higgs boson or Grand Unification of the forces of nature.

To move beyond the surface layers of reality, we need a story that connects the world we know with the deepest corners of the invisible world all around us. We cannot understand that hidden world with intuitions based solely on direct sensation. That is the story I want to tell here. I will take you on a journey to the heart of those mysteries that lie at the edge of our understanding of space, time, and the forces that operate within them. My goal is not to unnecessarily provoke or offend, but to prod you, just as we physicists ourselves have been prodded and dragged by new discoveries into a new reality that is at once both uncomfortable and uplifting.

Our most recent discoveries about nature's fundamental scales have chillingly altered our perception of the inevitability of our presence in the universe. They provide evidence too that the future will no doubt be radically different from what we might otherwise have imagined, and they too further decrease our cosmic significance.

We might prefer to deny this uncomfortable, inconvenient reality, this impersonal, apparently random universe, but if we view it in another context, all of this need not be depressing. A universe without purpose, which is the way it is as far as I can tell, is far more exciting than one designed just for us because it means that the possibilities of existence are so much more diverse and far ranging. How invigorating it is to find ourselves with an exotic menagerie to explore, with laws and phenomena that previously seemed beyond our wildest dreams, and to attempt to untangle the knotted confusion of experience and to search for some sense of order beneath. And how fascinating it is to discover that order, and to piece together a coherent picture of the universe on scales far beyond those that we may ever directly experience—a picture woven together by our ability to predict what will happen next, and the consequent ability to control the environment around us. How lucky to have our brief moment in the Sun. Every day that we discover something new and surprising, the story gets even better.

Part One

GENESIS

Chapter 1

FROM THE ARMOIRE TO THE CAVE

The simple inherit folly, but the prudent are crowned with knowledge.

—PROVERBS 14:18

*I*n my *beginning there was light.*

Surely there was light at the beginning of time, but before we can get to the beginning of time, we will need to explore our own beginnings, which also means exploring the beginning of science. And that means returning to the ultimate motive for both science and religion: the longing for *something else.* Something beyond the universe of our experience.

For many people, that longing translates into something that gives meaning and purpose to the universe and extends to a longing for some hidden place that is *better* than the world in which we live, where sins are forgiven, pain is absent, and death does not exist. Others, however, long for a hidden place of a very different sort, the physical world beyond our senses, the world that helps us understand how things behave the way they do, rather than why. This hidden world underlies what we experience, and the understanding of it gives us the power to change our lives, our environment, and our future.

9

The contrast between these two worlds is reflected in two very different works of literature.

The first, *The Lion, the Witch and the Wardrobe*, by C. S. Lewis, is a twentieth-century children's fantasy with decidedly religious overtones. It captures a childhood experience most of us have had—looking under the bed or in the closet or in the attic for hidden treasure or evidence that there is more out there than what we normally experience. In the book, several schoolchildren discover a strange new world, Narnia, by climbing into a large wardrobe in the country house outside London where they have been sequestered for their protection during the Second World War. The children help save Narnia with the aid of a lion, who lets himself be humiliated and sacrificed, Christlike, at an altar in order to conquer evil in his world.

While the religious allusion in Lewis's story is clear, we can also interpret it in another way—as an allegory, not for the existence of God or the devil, but rather for the remarkable and potentially terrifying possibilities of the unknown, possibilities that lie just beyond the edge of our senses, just waiting for us to be brave enough to seek them out. Possibilities that, once revealed, may enrich our understanding of ourselves or, for some who feel a need, provide a sense of value and purpose.

The portal to a hidden world inside the wardrobe is at once safe, with the familiar smell of oft-worn clothes, and mysterious. It implies the need to move beyond classical notions of space and time. For if nothing is revealed to an observer who is in front of or behind the wardrobe, and something is revealed only to someone inside, then the space experienced inside the wardrobe must be far larger than that seen from its outside.

Such a concept is characteristic of a universe in which space and time can be dynamical, as in the General Theory of Relativity, where, for example, from outside the "event horizon" of a black hole—that radius inside of which there is no escape—a black hole might appear to comprise a small volume, but for an observer inside (who has not yet been crushed to smithereens by the gravitational forces present), the volume can look quite different. Indeed, it is possible, though beyond the domain where we can perform reliable calculations, that the space inside a black hole might provide a portal to another universe disconnected from our own.

But the central point I want to return to is that the possibility of universes beyond our perception seems to be tied, in the literary and philosophical imagination, at least, to the possibility that space itself is not what it seems.

The harbinger of this notion, the "ur" story if you will, was written twenty-three centuries before Lewis penned his fantasy. I refer to Plato's *Republic*, and in particular to my favorite section, the Allegory of the Cave. But in spite of its early provenance, it illuminates more directly and more clearly both the potential necessity and the potential perils of searching for understanding beyond the reach of our immediate senses.

In the allegory, Plato likens our experience of reality to that of a group of individuals who live their entire lives imprisoned inside a cave, forced to face a blank wall. Their only view of the real world is that wall, which is illuminated by a fire behind them, and on which they see shadows moving. The shadows come from objects located behind them that the light of the fire projects on the wall.

I show the drawing below, which came from the high school text in which I first read this allegory, in a 1961 translation of Plato's dialogues.

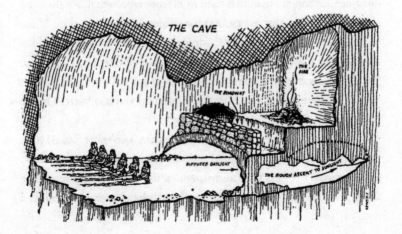

The drawing is amusing because it clearly reflects as much about the time it was drawn as it does the configuration of the cave described in the dialogue. Why, for example, are the prisoners here all women, and scantily clad ones at that? In Plato's day, any sexual allusion might easily have displayed young boys.

Plato argues that the prisoners will view the shadows *as* reality and even give them names. This is not unreasonable, and it is, in one sense, as we shall soon see, a very modern view of what reality is, namely that which we can directly measure. My favorite definition of reality still is that given by the science fiction writer Philip K. Dick, who said, "Reality is that which, when you stop believing in it, doesn't go away." For the prisoners, the shadows are what they see. They are also likely to hear only the echoes of noises made behind them as the sounds bounce off the wall.

Plato likened a philosopher to a prisoner who is freed from bondage and forced, almost against his will, to not only look at the fire, but to move past it, and out to the daylight beyond. First, the poor soul will be in distress, with the glare of the fire and the sunshine beyond the cave hurting his eyes. Objects will appear completely unfamiliar; they will not resemble their shadows. Plato argues that the new freeman may still imagine the shadows that he is used to as truer representations than the objects themselves that are casting the shadows.

If the individual is reluctantly dragged out into the sunshine, ultimately all of these sensations of confusion and pain will be multiplied. But eventually, he will become accustomed to the real world, will see the stars and Moon and sky, and his soul and mind will be liberated of the illusions that had earlier governed his life.

If the person returns to the cave, Plato argues, two things would happen. First, because his eyes would no longer be accustomed to the darkness, he would be less able to distinguish the shadows and recognize them, and his compatriots would view him as handicapped at best, and dim at worst. Second, he would no longer view the petty and myopic priorities of his former society, or the honors given to those who might best recognize the shadows and predict their future, as worthy of his respect. As Plato poetically put it, quoting from Homer:

"Better to be the poor servant of a poor master, and to endure anything, rather than think as they do and live after their manner."

So much for those whose lives are lived entirely in illusion, which Plato suggests includes most of humanity.

Then, the allegory states that the journey upward—into the light—is the ascent of the soul into the intellectual world.

Clearly in Plato's mind only a retreat to the purely "intellectual world," a journey reserved for the few—aka philosophers—could replace illusion with reality. Happily, that journey is far more accessible today using the techniques of science, which combine reason and reflection with empirical inquiry. Nevertheless, the same challenge remains for scientists today: to see what is behind the shadows, to see that which, when you drop your preconceptions, doesn't disappear.

While Plato doesn't explicitly mention it, not only would his fellow prisoners view the poor soul who had ventured out and returned as handicapped, but they would likely think he was crazy if he talked about the wonders that he had glimpsed: the Sun, the Moon, lakes, trees, and other people and their civilizations.

This idea is strikingly modern. As the frontiers of science have moved further and further away from the world of the familiar and the world of common sense as inferred from our direct experience, our picture of the reality underlying our experience is getting increasingly difficult for us to comprehend or accept. Some find it more comforting to retreat to myth and superstition for guidance.

But, we have every reason to expect that "common sense," which first evolved to help us cope with predators in the savannas of Africa, might lead us astray when we attempt to think about nature on vastly different scales. We didn't evolve to intuitively understand the world of the very small, the very big, or the very fast. We shouldn't expect the rules we have come to rely on for our daily lives to be universal. While that myopia was useful from an evolutionary perspective, as thinking beings we can move beyond it.

In this regard, I cannot resist quoting one last admonition in Plato's allegory:

"In the world of knowledge the idea of good appears last of all and is seen only with an effort; and, when seen, is also inferred to be the author of all things good and right, parent of light, and . . . the immediate source of reason and truth."

Plato further argues that this is what those who would act rationally should strive for, in both public and private life—seeking the "good" by focusing on reason and truth. He suggests that we can only do so by exploring the realities that underlie the world of our direct experience, rather than by exploring the illusions of a reality that we might *want* to exist. Only through rational examination of what is real, and not by faith alone, is rational action—or good—possible.

Today, Plato's vision of "pure thought" has been replaced by the scientific method, which, based on both reason *and* experiment, allows us to discover the underlying realities of the world. Rational action in public and private life now requires a basis in both reason and empirical investigation, and it often requires a departure from the solipsistic world of our direct experience. This principle is the source of most of my own public activism in opposition to government policies based on ideology rather than evidence, and it is also probably why I respond so negatively to the concept of the "sacred"—implying as it does some idea or admonition that is off-limits to public questioning, exploration, discussion, and sometimes ridicule.

It is hard to state this view more strongly than I did in a *New Yorker* piece: "Whenever scientific claims are presented as unquestionable, they undermine science. Similarly, when religious actions or claims about sanctity can be made with impunity in our society, we undermine the basis of modern secular democracy. We owe it to ourselves and to our children not to give a free pass to governments—totalitarian, theocratic, or democratic—that endorse, encourage, enforce, or otherwise legitimize the suppression of open questioning in order to protect ideas that are considered 'sacred.' Five hundred years of science have liberated humanity from the shackles of enforced ignorance."

Philosophical reflections aside, the prime reason I am introducing Plato's cave here is that it can provide a concrete example of the nature of the scientific discoveries at the heart of the story I want to tell.

Imagine a shadow that our prisoners might see on the wall, displayed by an evil puppeteer located on a ledge in front of the fire:

This shadow displays both length and directionality, two concepts that we, who are not confined to the cave, take for granted.

However, as the prisoners watch, this shadow changes:

Later it looks like this:

And again later like this:

And later still, like this:

What would the prisoners infer from all of this? Presumably, that concepts such as length or direction have no absolute meaning. The objects in their world can change both length and directionality arbitrarily. In the reality of their direct experience, neither length nor directionality appears to have significance.

What will the natural philosopher, who has escaped to the surface to explore the richer world beyond the shadows, discover? He will see that the shadow is first of all just a shadow: a two-dimensional image on the wall cast from a real, three-dimensional object located behind the prisoners. He will see that the object has a fixed length that never changes, and that it's accompanied by an arrow that is always on the same side of the object. From a vantage point slightly above the object, he sees that the series of images results from the projection of a rotating weather vane onto the wall:

When he returns to join his former colleagues, the philosopher-scientist can explain that an absolute quantity called length *doesn't* change over time, and that directionality can be assigned unambiguously to certain objects as well. He will tell his friends that the real world is three-dimensional, not two-dimensional, and that once they understand, all of their confusion about the seemingly arbitrary changes will disappear.

Would they believe him? It would be a tough sell because they won't have an intuitive idea of what a rotation is (after all, with an intuition based purely on two-dimensional experience, it would likely be difficult to "picture" mentally any rotations in a third dimension). Blank stares? Probably. The loony bin? Maybe. However, he might win over the community by stressing attractive characteristics associated with his claim: *behavior that on the surface appears to be complex and arbitrary can be shown to result from a much simpler underlying picture of nature, and seemingly disparate phenomena are actually connected and can be part of a unified whole.*

Better still, he could make predictions that his friends could test. First, he could argue that, if the apparent change in length of the shadows measured by the group is really due to a rotation in a third dimension, whenever the length of the object briefly vanishes, it will immediately reemerge with the arrow pointing in the opposite direction. Second, he could argue that as the length oscillates, the maximum length of the shadow when the arrow is pointing in one direction will always be exactly the same as the maximum length of the shadow when it is pointing in the other direction.

Plato's cave thus becomes an allegory for far more than he may have intended. Plato's freed man discovers the hallmarks of the remarkable true story of our own struggle to understand nature on its most fundamental scales of space, time, and matter. We too have had to escape the shackles of our prior experience to uncover profound and beautiful simplifications and predictions that can be as terrifying as they are wonderful.

But just as the light beyond Plato's cave is painful to the eyes at first, with time it becomes mesmerizing. And once witnessed, there is no going back.

Chapter 2

SEEING IN THE DARK

Let there be light: and there was light.
—GENESIS 1:3

In the beginning there was light.

It is no coincidence that the ancients imagined in Genesis that light was created on the first day. Without light, there would be little awareness of the vast universe surrounding us. When we nod and say, "I see," to a friend who is trying to explain something, we convey far more than just an observation, but rather a fundamental understanding.

Plato's allegory was appropriately centered on light—light from a fire to cast the shadows on the cave wall and light from the outside to temporarily blind the freed prisoner and then illuminate the real world for him. Like the prisoners in the cave, we too are prisoners of light— almost everything we learn about the world we learn from what we see.

While the most significant words in the Western religious canon may be *Let there be light*, in the modern world this phrase now has a completely different significance from what it once did. Human beings may be prisoners of light, but so is the universe. What once appeared as a whim of a Judeo-Christian God, or other gods before that one, we now understand to be required by the very laws that allow both heaven, and more important, Earth, to exist. You cannot have one without the other. Earth, or matter, follows light.

This change in perception underlies almost every development in the edifice we call modern science. I am writing these words as I stare out from a ship at one of the Galápagos Islands, which Charles Darwin made famous, and which made him famous in return, as he changed our perception of life and its diversity with a single brilliant realization: that all living species developed through the natural selection of small inherited variations that are passed along to future generations by survivors. As surely as the understanding of evolution changed everything about our understanding of biology, our changing understanding of light changed everything about our physical understanding of our place in the universe. As a useful fringe benefit, this change resulted in virtually all of the technology on which the modern world is based.

The extent to which our observations of the world imprison our minds, and frame our description of the fabric of the universe, remained unappreciated for more than twenty centuries following Plato. Once serious minds began to investigate in detail the hidden nature of the universe, it took over four centuries for them to fully resolve the question What is light?

Perhaps the most serious modern mind, although certainly not the first, to ask this question was also one of the most famous—and oddest—scientists in history: Isaac Newton. It is not inappropriate to classify Newton as a modern mind—after all, his seventeenth-century *Principia: Mathematical Principles of Natural Philosophy* uncovered the classical laws of motion and laid the basis for his theory of gravity, both of which form the foundation of much of modern physics. Nevertheless, as John Maynard Keynes pointed out:

Newton was not the first of the age of reason, he was the last of the magicians, the last of the Babylonians and Sumerians, the last great mind that looked out on the visible and intellectual world

with the same eyes as those who began to build our intellectual
inheritance rather less than 10,000 years ago.

The truth of this statement reflects the revolutionary importance
of Newton's work. After the *Principia*, no rational person could view
the world the same way the ancients had viewed it. But it also reflects
the character of Newton himself. He devoted far more time, and far
more ink, to writing about the occult, alchemy, and searching for hidden
meanings and codes in the Bible—focusing in particular on the Book of
Revelation and mysteries associated with the ancient Temple of Solo-
mon—than he did to writing about physics.

Newton was also one in a long line of people, which extends before
and after him, who felt that he had been specifically chosen by God to
help reveal the true meaning of the Scriptures. To what extent his studies of the universe derived from his fascination with the Bible is not
clear, but it does seem reasonable to conclude that his primary interest
was in theology, and that natural philosophy came in well below that,
and probably below alchemy as well.

Many individuals point to Newton's fascination with God as evidence of the compatibility between science and religion, and to assert
that modern science owes its existence to Christianity. This confuses
history with causality. It is undeniable that many of the early giants
of modern Western natural philosophy, from Newton onward, were
deeply religious, although Darwin lost much, if not all, of his religious belief later in life. But remember that during much of this period there were primarily two sources of education and wealth: the
Church and the Crown. The Church was the National Science Foundation of the fifteenth, sixteenth, and seventeenth centuries. All institutions of higher learning were tied to various denominations, and
it was unthinkable for any educated person to not be affiliated with
the Church. And as Giordano Bruno and later Galileo discovered, it
was unpleasant at best to counter its doctrine. It would have been

remarkable for any of these leading early scientific thinkers to have been anything but religious.

The religiosity of the early scientific pioneers is also cited today by sophists who claim that science and religious doctrine are compatible, but who confuse science and scientists. In spite of frequent appearances to the contrary, scientists are people. And like all people they are capable of holding many potentially mutually contradictory notions in their head at the same time. No correlation between divergent views held by any individual is representative of anything but human foibles.

To claim that some scientists are or were religious is like saying some scientists are Republicans or some are flat-earthers or some are creationists. It doesn't imply causality or consistency. My friend Richard Dawkins has told me of a professor of astrophysics who, during the day, writes papers that are published in astronomical journals assuming that the universe is more than 13 billion years old, but then goes home and privately espouses the literal biblical claim that the universe is six thousand years old.

What determines intellectual consistency or lack thereof in the sciences is a combination of rational arguments with subsequent evidence and continued testing. It is perfectly reasonable to claim that religion, in the Western world, may be the mother of science. But as any parent knows, children rarely grow up to be models of their parents.

Newton may, following tradition, have been motivated to look at light because it was a gift from God. But we remember his work not because of his motivation, but because of what he discovered.

Newton was convinced that light was made of particles, which he referred to as corpuscles, while Descartes, and later Newton's nemesis Robert Hooke, and still later the Dutch scientist Christiaan Huygens, all claimed that light was a wave. One of the key observations that appeared to support the wave theory was that white light, such as light from the Sun, could split into all the colors of the rainbow when passed through a prism.

As was often the case during his life, Newton believed that he was correct and several of his most famous contemporaries (and competitors) were wrong. To demonstrate this, he devised a clever experiment using prisms that he first performed while at home in Woolsthorpe, to escape the bubonic plague ravaging Cambridge. As he reported at the Royal Society in 1672, on the forty-fourth try, he observed precisely what he hoped he would see.

Advocates of the wave theory had argued that light waves were made of white light and that the light split into colors when it passed through a prism because of "corruption" of the rays as they traversed the glass. In this case, the more glass, the more splitting.

Newton reasoned that this was not the case, but that light is made of colored particles that combine together to appear white. (With a nod to his occult fascination, Newton classified the colored particles of the *spectrum*—a term he coined—into seven different types: red, orange, yellow, green, blue, indigo, and violet. From the time of the Greeks, the number seven had been considered to possess mystical qualities.) To demonstrate that the wave/corruption picture was incorrect, Newton passed a beam of white light through two prisms held in opposite orientations. The first prism split the light into its spectrum, and the second recomposed it back into a single white light beam. This result would have been impossible if the glass had corrupted the light. A second prism would have simply made the situation worse and would not have caused the light to revert back to its original state.

This result does not in fact disprove the wave theory of light (it actually supports it, because light slows down as it bends upon entering the prism, just as waves would do). But since the advocates of that theory had argued (incorrectly) that the spectral splitting was due to corruption, Newton's demonstration that this was not the case struck a significant blow in favor of his particle model.

Newton went on to discover many other facets of light that we use today in our understanding of the wave nature of light. He showed that every color of light has a unique bend angle when passing through a glass prism. He also showed that all objects appear to be the same color as the color of the light beam illuminating them. And he showed that colored light will not change its color no matter how many times it is reflected by or passes through a prism.

All of these results, including his original result, can be explained simply if white light is indeed composed of a collection of different colors—that much he got right. But they can't be explained if light is made of different-colored particles. Rather, white light is composed of waves of many different wavelengths.

Newton's opponents did not give up easily, even in the face of Newton's rising popularity and the death of his chief opponent, Hooke. They did not give up even after Newton's election as president of the Royal Society in 1703, the year before he actually published his research on light in his epic *Opticks*. Indeed, the debate on the nature of light continued to rage on for over a century.

Part of the problem with a wave picture of light was the question "What is it that light is a wave *of*, exactly?" And if it is a wave, then since all known waves require some medium, what medium does it travel in? These questions were sufficiently perplexing that practitioners of the wave theory had to resurrect a new invisible substance permeating all space, the ether.

The resolution of this conundrum came, as such resolutions often do, from a totally unexpected corner of the physical world, one full of sparks, and spinning wheels.

When I was a young professor at Yale—in the ancient but huge office I was lucky enough to commandeer when an equally ancient colleague retired—there was left hanging for me a copy of a photograph of Michael Faraday taken in 1861. I have treasured it ever since.

I don't believe in hero worship, but if I did, Faraday would be up there with the best. Perhaps more than any other scientist of the nineteenth century, he is responsible for the technology that powers our current civilization. Yet he had little formal education and at age fourteen became a bookbinder's apprentice. Later in his career, after achieving world recognition for his scientific contributions, he insisted on keeping to his humble roots, turning down a knighthood and twice turning down the presidency of the Royal Society. Later on he refused to advise the British government on the production of chemical weapons for use in the Crimean War, citing ethical reasons. And for more than thirty-three years he gave a series of Christmas lectures at the Royal Institution to excite young people about science. What's not to like?

Much as one might admire the man, it is the scientist who matters here for our story. Faraday's first scientific lesson is one I tell my students: always suck up to your professors. At the age of twenty, after completing seven years of apprenticeship as a bookbinder, Faraday attended the lectures of the famous chemist Humphry Davy, then the head of the Royal Institution. Afterward Faraday presented Davy with a three-hundred-page, beautifully bound book containing the notes Faraday had taken during the lectures. Within a year, Faraday was appointed Davy's secretary and shortly thereafter got an appointment as chemical assistant in the Royal Institution. Later on, Faraday learned the same lesson but with the opposite result. Following his excitement over some early, quite significant experiments that he performed, Faraday accidentally forgot to acknowledge work with Davy in his published results. This accidental snub probably resulted in his being reassigned to other activities by Davy and delaying his world-changing research by several years.

When reassigned, Faraday had been working on the "hot" area of scientific research, the newly discovered connections between electricity and magnetism, driven by results of the Danish physicist Hans Christian Oersted. These two forces seem quite different, yet have odd similarities. Electric charges can attract or repel. So can magnets. Yet magnets always seem to have two poles, north and south, which cannot be isolated, while electric charges can individually be positive or negative.

For some time, scientists and natural philosophers had wondered if the two forces might have some hidden connection, and the first empirical clue came to Oersted by accident. In 1820, while delivering a lecture, Oersted saw that a compass needle was deflected when an electric current from a battery was switched on. A few months later he followed up on this observation and discovered that a current of moving electric charges, which we now commonly call an electric current, produced a magnetic attraction that caused compass needles to point in a circle around the wire.

He had blazed a new trail. Word spread quickly among scientists, through the Continent and across the English Channel. Moving electric charges produced a magnetic force. Could there be other connections? Could magnets in turn influence electric charges?

Scientists searched for such a possibility, without success. Davy and another colleague tried to build an electric motor based on the connection discovered by Oersted, but failed. Faraday ultimately got a wire with a current in it to move around a magnet, which did form a crude sort of motor. It was this exciting development that he reported without citing Davy's name.

Partly this was mere gamesmanship. No new fundamental phenomenon was being uncovered. Perhaps this was the rationale for one of my favorite (likely apocryphal) stories about Faraday. It is said that William Gladstone, later to be British prime minister, heard of Faraday's laboratory, full of weird devices, and asked in 1850 what the practical value of all this study into electricity was. Faraday was purported to have replied, "Why, sir, there is every probability that you will soon be able to tax it."

Apocryphal or not, both great irony and truth are in that witty comeback. Curiosity-driven research may seem self-indulgent and far from the immediate public good. However, essentially all of our current quality of life, for people living in the first world, has arisen from the fruits of such research, including all the electric power that drives almost every device we use.

Two years after Davy's death in 1829, and six years after Faraday had become director of the laboratory of the Royal Institution, he made the discovery that cemented his reputation as perhaps the greatest experimental physicist of the nineteenth century—magnetic induction. Since 1824, he had tried to see if magnetism could alter the current flowing in a nearby wire or otherwise produce some kind of electric force on charged particles. He primarily wanted to see if magnetism could induce electricity, just as Oersted had shown that electricity, and electric currents in particular, could produce magnetism.

On October 28, 1831, Faraday recorded in his laboratory notebook a remarkable observation. While closing the switch to turn on a current in a wire wound around an iron ring to magnetize the iron, he noticed a current flow momentarily in another wire wrapped around the same iron ring. Clearly the mere presence of a nearby magnet could not cause an electric current to flow in a wire—but turning the magnet on or off could. Subsequently he showed that the same effect occurred if he moved a magnet near a wire. As the magnet came closer or moved away, a current would flow in the wire. Just as a moving charge created a magnet, somehow a moving magnet—or a magnet of changing strength—created an electric force in the nearby wire and produced a current.

If the profound theoretical implication of this simple and surprising result is not immediately apparent, you can be forgiven, because the implication is subtle, and it took the greatest theoretical mind of the nineteenth century to unravel it.

To properly frame it, we need a concept that Faraday himself introduced. Faraday had little formal schooling and was largely self-taught and thus was never comfortable with mathematics. In another probably apocryphal story, Faraday boasted of using a mathematical equation only one time in all of his publications. Certainly, he never described the important discovery of magnetic induction in mathematical terms.

Because of his lack of comfort with formal mathematics, Faraday was forced to think in pictures to gain intuition about the physics behind his observations. As a result he invented an idea that forms the cornerstone of all modern physics theory and resolved a conundrum that had puzzled Newton until the end of his days.

Faraday asked himself, How does one electric charge "know" how to respond to the presence of another, distant electric charge? The same question had been posed by Newton in terms of gravity, where he earlier wondered how the Earth "knew" to respond as it did to the gravitational pull of the Sun. How was the gravitational force conveyed from one body to another? To this, he gave his famous response "Hypotheses *non fingo*," "I frame no hypotheses," suggesting that he had worked out the force law of gravity and showed that his predictions matched observations, and that was good enough. Many of us physicists have subsequently used this defense when asked to explain various strange physics results—especially in quantum mechanics, where the mathematics works, but the physical picture often seems crazy.

Faraday imagined that each electric charge would be surrounded by an electric "field," which he could picture in his head. He saw the field as a bunch of lines emanating radially outward from the charge. The field lines would have arrows on them, pointing outward if the charge was positive, and inward if it was negative:

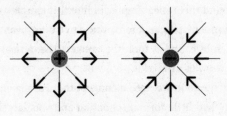

He further imagined that the number of field lines increased as the magnitude of the charge increased:

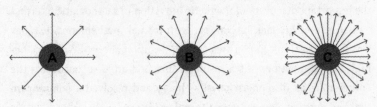

The utility of this mental picture was that Faraday could now intuitively understand both what would happen when another test charge was put near the first charge and why. (Whenever I use the colloquial *why*, I mean "how.") The test charge would feel the "field" of the first charge wherever the second charge was located, with the strength of the force being proportional to the number of field lines in the region, and the direction of the force being along the direction of the field lines. Thus, for example, the test charge in question would be pushed outward in the direction shown:

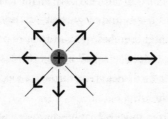

One can do more than this with Faraday's pictures. Imagine placing two charges near each other. Since field lines begin at a positive charge and end on a negative charge and can never cross, it is almost intuitive that the field lines in between two positive charges should appear to repel each other and be pushed apart, whereas between a positive and a negative charge they should connect together:

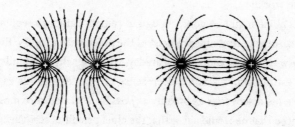

Once again, if a test charge is placed anywhere near these two charges, it would feel a force in the direction of the field lines, with a strength proportional to the number of field lines in that region.

Faraday thus pictured the nature of electric forces between particles in a way that would otherwise require solving the algebraic equations that describe electrical forces. What is most amazing about these pictures is that they capture the mathematics exactly, not merely approximately.

A similar pictorial view could be applied to magnets, and magnetic fields, reproducing the magnetic force law between magnets, experimentally verified by Coulomb, or current-carrying wires, derived by André-Marie Ampere. (Up until Faraday, all the heavy lifting in discovering the laws of electricity and magnetism was done by the French.)

Using these mental crutches, we can then reexpress Faraday's discovery of magnetic induction as follows: an increase or decrease in the number of magnetic field lines going through a loop of wire will cause a current to flow in the wire.

Faraday recognized quickly that his discovery would allow the conversion of mechanical power into electrical power. If a loop of wire was attached to a blade that was made to rotate by, say, a flow of water, such as a waterwheel, and the whole thing was surrounded by a magnet, then as the blade turned the number of magnetic field lines going through the wire would continuously change, and a current would continuously be generated in the wire. Voilà, Niagara Falls, hydroelectricity, and the modern world!

This alone might be good enough to cement Faraday's reputation as the greatest experimental physicist of the nineteenth century. But technology wasn't what motivated Faraday, which is why he stands so tall in my estimation; it was his deep sense of wonder and his eagerness to share his discoveries as broadly as possible that I admire most. I am convinced that he would agree that the chief benefit of science lies in its impact in changing our fundamental understanding of our place in the cosmos. And ultimately, this is what he did.

I cannot help but be reminded of another more recent great experimental physicist, Robert R. Wilson—who, at age twenty-nine, was head of the Research Division at Los Alamos, which developed the atomic bomb during the Manhattan Project. Many years later he was the first director of the Fermi National Accelerator Laboratory in Batavia, Illinois. When Fermilab was being built, in 1969 Wilson was summoned before Congress to defend the expenditure of significant funds on this exotic new accelerator, which was to study the fundamental interactions of elementary particles. Asked if it contributed to national security (which would have easily justified the expenditure in the eyes of the congressional committee members), he bravely said no. Rather:

> *It only has to do with the respect with which we regard one another, the dignity of men, our love of culture. . . . It has to do with, are we good painters, good sculptors, great poets? I mean all the things that we really venerate and honor in our country and are patriotic about. In that sense, this new knowledge has all to do with honor and country, but it has nothing to do directly with defending our country except to help make it worth defending.*

Faraday's discoveries allowed us to power and create our civilization, to light up our cities and our streets, and to run our electric devices. It is hard to imagine any discovery that is more deeply ingrained in the workings of modern society. But more deeply, what makes his contribution to our story so remarkable is that he discovered a missing piece of the puzzle that changed the way we think about virtually everything in the physical world today, starting with light itself. If Newton was the last of the magicians, Faraday was the last of the modern scientists to live in the dark, regarding light. After his work, the key to uncovering the true nature of our main window on the world lay in the open waiting for the right person to find it.

· · ·

Within a decade, a young Scottish theoretical physicist, down on his luck, took the next step.

Chapter 3

THROUGH A GLASS, LIGHTLY

*Nothing is too wonderful to be true, if it be consistent
with the laws of nature; and in such things as these,
experiment is the best test of such consistency.*

—FARADAY, LABORATORY JOURNAL ENTRY #10,040

(MARCH 18, 1849)

The greatest theoretical physicist of the nineteenth century, James Clerk Maxwell, whom Einstein would later compare to Newton for his impact on physics, was coincidentally born in the same year that Michael Faraday made his great experimental discovery of induction.

Like Newton, Maxwell also began his scientific career fascinated by color and light. Newton had explored the spectrum of visible colors into which white light splits when traversing a prism, but Maxwell, while still a student, investigated the reverse question: What is the minimal combination of primary colors that would reproduce for human perception all the visible colors contained in white light? Using a collection of colored spinning tops, he demonstrated that essentially all colors we perceive can result from mixtures of red, green, and blue—a fact familiar to anyone who has plugged RGB cables into a color television. Maxwell used this realization to produce the world's first, rudimentary color photograph. Later he became fascinated with polarized light, which results from light waves whose electric and magnetic fields oscillate only in certain directions. He sandwiched blocks of gelatin between polarizing prisms and shined light through them. If the two prisms allowed only light to pass that was polarized in different perpendicular directions, then if one was placed behind the other, no light would make it through. However, if stresses were present in the gelatin, then the light could have its axis of polarization rotated as it passed through the material, so that some light might then make it through the second prism. By searching for such fringes of light passing through the second prism, Maxwell could explore for stresses in the material. This has become a useful tool today for exploring possible material stresses in complex structures.

Even these ingenious experiments do not adequately represent the power of Maxwell's voracious intellect or his mathematical ability, which were both manifest at a remarkably early age. Tragically, Maxwell died at the age of forty-eight and had precious little time to accomplish all that he did. His inquisitive nature was reflected in a passage his mother added to a letter from his father to his sister-in-law when Maxwell was only three:

> *He is a very happy man, and has improved much since the weather got moderate; he has great work with doors, locks, keys, etc., and "show me how it doos" is ever out of his mouth. He also investigates the hidden course of streams and bell-wires, the way the water gets from the pond through the wall.*

After his mother's untimely death (of stomach cancer, to which Maxwell would later succumb at the same age), his education was interrupted, but by the age of thirteen he had hit his stride at the prestigious Edinburgh Academy, where he won the prize for mathematics, and also for English and poetry. He then published his first scientific paper—concerning the properties of mathematical curves—which was presented at the Royal Society of Edinburgh when he was only fourteen.

After this precocious start, Maxwell thrived at university. He graduated from Cambridge, becoming a fellow of the college within a year after graduation, which was far sooner than average for most graduates. He left shortly thereafter and returned to his native Scotland to take up a chair in natural philosophy in Aberdeen.

At only twenty-five, he was head of a department and teaching fifteen hours a week plus an extra free lecture for a nearby college for working men (something that would be unheard of for a chaired professor today, and something that I find difficult to imagine doing myself and still having any energy left for research). Yet Maxwell nevertheless found time to solve a problem that was two centuries old: How could Saturn's rings remain stable? He concluded that the rings must be made of small particles, which garnered him a major prize that had been set up to encourage an answer to this question. His theory was confirmed more than a hundred years later when *Voyager* provided the first close-up view of the planet.

You would think that, after his remarkable output, he would have been able to remain secure in his professorship. However, in 1860, the same year that he was awarded the Royal Society's prestigious Rumford Medal for his work on color, the college where he lectured merged with another college and had no room for two professors of natural philosophy. In what must surely go down in history as one of the dumbest academic decisions ever made (and that is a tough list to top), Maxwell was unceremoniously laid off. He tried to get a chair in Edinburgh, but again the position was given to another candidate. Finally, he found a position down south, at King's College, London.

One might expect Maxwell to have been depressed or disconsolate because of these developments, but if he was, his work reflected no signs of it. The next five years at King's were the most productive period in his life. During this time he changed the world—four times.

The first three contributions were the development of the first light-fast color photograph; the development of the theory of how particles in a gas behave (which helped establish the foundations of the field now known as statistical mechanics—essential for understanding the properties of matter and radiation); and finally his development of "dimensional analysis," which is perhaps the tool most frequently used by modern physicists to establish deep relationships between physical quantities. I just used it last year, for example, with my colleague Frank Wilczek, to demonstrate a fundamental property of gravity relevant to understanding the creation of our universe.

Each contribution on its own would have firmly established Maxwell among the greatest physicists of his day. However, his fourth contribution ultimately changed everything, including our notions of space and time.

During his period at King's, Maxwell frequented the Royal Institution, where he came in contact with Michael Faraday, who was forty years older but still inspirational. Perhaps these meetings encouraged Maxwell to return his focus to the exciting developments in electricity and magnetism, a subject he had begun to investigate five years earlier. Maxwell used his considerable mathematical talents to describe and understand the phenomena explored by Faraday. He began by putting Faraday's hypothesized lines of force on a firmer mathematical footing, which allowed him to explore in more depth Faraday's discovery of induction. Over the dozen years between 1861 and 1873, Maxwell put the final touches on his greatest work, a complete theory of electricity and magnetism.

To do this, Maxwell used Faraday's discovery as the key to revealing that the relationship between electricity and magnetism is symmetrical. Oersted's and Faraday's experiments had shown, simply, that a current of moving charges produces a magnetic field; and that a changing magnetic field (produced by moving a magnet or simply turning on a current to produce a magnet) produces an electric field.

Maxwell first expressed these results mathematically in 1861, but soon realized that his equations were incomplete. Magnetism appeared to be different from electricity. Moving charges create a magnetic field, but a magnetic field can create an electric field even without moving—just by changing. As Faraday discovered, turning on a current, which produces a changing magnetic field as the current ramps up, produces an electric force that causes a current to flow in another nearby wire.

Maxwell recognized that to make a complete and consistent set of equations for electricity and magnetism he had to add an extra term to the equations, representing what he called a "displacement current." He reasoned that moving charges, namely a current, produce a magnetic field, and moving charges represent one way to produce a changing electric field (since the field from each charge changes in space as the charge moves along). So, maybe, a changing electric field—one that gets stronger or weaker—in a region with no charges in motion, could produce a magnetic field.

Maxwell envisioned that if he hooked up two parallel plates to opposite poles of a battery, each plate would get charged with an opposite charge as current flowed from the battery. This would produce a growing electric field between the plates and would also produce a magnetic field around the wires connected to the plates. For his equations to be completely consistent, Maxwell realized, the increasing electric field between the plates should also produce a magnetic field in that empty space between the plates. And that field would be the same as any magnetic field produced by a real current flowing through that space between the plates.

So Maxwell altered his equations by adding a new term (displacement current) to produce mathematical consistency. This term effectively behaved like an imaginary current, flowing between the plates producing a changing electric field identical in magnitude to the actual changing electric field in the empty space between the plates. It also was the same as the magnetic field that a real current would produce if it flowed between the plates. Such a magnetic field does in fact arise when you perform the experiment with parallel plates, as undergraduates demonstrate every day in physics laboratories around the world.

Mathematical consistency and sound physical intuition generally pay off in physics. This subtle change in the equations may not seem like much, but its physical impact is profound. Once you remove real electric charges from the picture, it means that you can describe everything about electricity and magnetism entirely in terms of the hypothetical "fields" that Faraday had relied upon purely as a mental crutch. The connections between electricity and magnetism can thus be simply stated: A changing electric field produces a magnetic field. A changing magnetic field produces an electric field.

Suddenly the fields appear in the equations as real physical objects in their own right and not merely as a way to quantify the force between charges. Electricity and magnetism became inseparable. It is impossible to talk about electrical forces alone because, as I will shortly show, one person's electric force is another person's magnetic force, depending on the circumstances of the observer, and whether the field is changing in his frame of reference.

We now refer to *electromagnetism* to describe these phenomena, for a good reason. After Maxwell, electricity and magnetism were no longer viewed as separate forces of nature. They were different manifestations of one and the same force.

Maxwell published his complete set of equations in 1865 and later simplified them in his textbook of 1873. These would become famous as the four Maxwell's Equations, which (admittedly rewritten in modern mathematical language) adorn the T-shirts of physics undergraduates around the world today. We can thus label 1873 as establishing the second great unification in physics, the first being Newton's recognition that the same force governed the motion of celestial bodies as governed falling apples on Earth. Begun with Oersted's and Faraday's experimental discoveries, this towering achievement of the human intellect was completed by Maxwell, a mild-mannered young theoretical physicist from Scotland, exiled to England by the vicissitudes of academia.

Gaining a new perspective on the cosmos is always—or should be—immensely satisfying. But science adds an additional and powerful benefit. New understanding also breeds tangible and testable consequences, and often immediately.

So it was with Maxwell's unification, which now made Faraday's hypothetical fields literally as real as the nose on your face. *Literally*, because it turns out you couldn't see the nose on your face without them.

Maxwell's genius didn't end just with codifying the principles of electromagnetism in elegant mathematical form. He used the mathematics to unravel the hidden nature of that most fundamental of all physical quantities—which had eluded the great natural philosophers from Plato to Newton. The most observable thing in nature: light.

Consider the following thought experiment. Take an electrically charged object and jiggle it up and down. What happens as you do this?

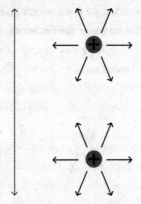

Well, an electric field surrounds the charge, and when you move the charge, the position of the field lines changes. But, according to Maxwell, this changing electric field will produce a magnetic field, which will point in and out of the paper as shown below:

Here the field line pointing into the paper has a cross (the back of an arrow), and that pointing out of the paper has a dot (the tip of an arrow). This field will flip direction as the charge changes the direction of its motion from upward to downward.

But we should not stop there. If I keep jiggling the charged object, the electric field will keep changing, and so will the induced magnetic field. But a changing magnetic field will produce an electric field. Thus there are new induced electric field lines, which point vertically, changing from up to down as the magnetic field reverses its sign. I display the electric field line to the right only for lack of space, but the mirror image will be induced on the left-hand side.

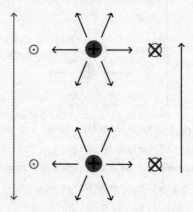

But that changing electric field will in turn produce a changing magnetic field, which would exist farther out to the right and left of the diagram, and so on.

Jiggling a charge produces a succession of disturbances in both electric and magnetic fields that propagate outward, with the change in each field acting as a source for the other, due to the rules of electromagnetism as Maxwell defined them. We can extend the picture shown above to a 3-D image that captures the full nature of the changing as shown below:

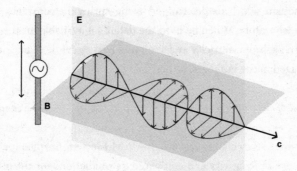

We see a wave of electric and magnetic disturbances, namely an electromagnetic wave moving outward, with electric and magnetic fields oscillating in space, and time, and with the two fields oscillating in directions that are perpendicular to each other and also the direction of the wave.

Even before Maxwell had written down the final form of his equations, he showed that oscillating charges would produce an electromagnetic wave. But he did something far more significant. He calculated the speed of that wave, in a beautiful and simple calculation that is probably my favorite derivation to show undergraduates. Here it is:

We can quantify the strength of an electric force by measuring its magnitude between two charges whose magnitude we already know. The force is proportional to the product of the charges. Let's call the constant of proportionality A.

Similarly we can quantify the strength of the magnetic force between two electromagnets, each with a current of known magnitude. This force is proportional to the product of the currents. Let's call the constant of proportionality in this case B.

Maxwell showed that the speed of an electromagnetic disturbance that emanates from an oscillating charge can be rendered precisely in terms of the measured strength of electricity and the measured strength of magnetism, which are determined by measuring the constants A and B in the laboratory. When he used the data then available for the measured strength of electricity and the measured strength of magnetism and plugged in the numbers, he derived:

Speed of electromagnetic waves ≈ 311,000,000 meters per second

A famous story claims that when Albert Einstein finished his General Theory of Relativity and compared its predictions for the orbit of Mercury to the measured numbers, he had heart palpitations. One can only imagine, then, the excitement that Maxwell must have had when he performed his calculation. For this number, which may seem arbitrary, was well known to him as the speed of light. In 1849, the French physicist Fizeau had measured the speed of light, an extremely difficult measurement back then, and had obtained:

Speed of light ≈ 313,000,000 meters per second

Given the accuracy available at the time, these two numbers are identical. (We now know this number far more precisely as 299,792,458 meters per second, which is a key part of the modern definition of the meter.)

In his typical understated tone, Maxwell noted in 1862, when he first performed the calculation, "We can scarcely avoid the conclusion that light consists in the transverse undulations of the same medium which is the cause of electric and magnetic phenomena."

In other words, light is an electromagnetic wave.

Two years later, when he finally wrote his classic paper on electromagnetism, he added somewhat more confidently, "Light is an electromagnetic disturbance propagated through the field according to electromagnetic laws."

With these words, Maxwell appeared to have finally put to rest the two-thousand-year-old mystery regarding the nature and origin of light. His result came, as great insights often do, as an unintended by-product of other fundamental investigations. In this case, it was a by-product of one of the most important theoretical advances in history, the unification of electricity and magnetism into a single beautiful mathematical theory.

. . .

Before Maxwell, the chief source of wisdom came from a faith in divinity via Genesis. Even Newton relied upon this source for understanding the origin of light. After 1862, however, everything changed.

James Clerk Maxwell was deeply religious, and like Newton before him, his faith sometimes led him to make strange assertions about nature. Nevertheless, like the mythical character Prometheus before him, who stole fire from the gods and gave it to humans to use as a tool to forever change their civilization, so too Maxwell stole fire from the Judeo-Christian God's first words and forever changed their meaning. Since 1873, generations of physics students have proudly proclaimed:

"Maxwell wrote down his four equations and said, *Let there be light!*"

Chapter 4

THERE, AND BACK AGAIN

He set the earth on its foundations; it can never
be moved.

—PSALMS 104:5

When Galileo Galilei was being tried in 1633 for heresy for "holding as true the false doctrine taught by some that the Sun is the center of the world," he allegedly muttered under his breath in front of his Church inquisitors, "And yet it moves." With these words, his revolutionary nature once again sprang forth, in spite of his having been forced to publicly adhere to the archaic position that the Earth was fixed.

While the Vatican eventually capitulated on Earth's motion, the poor God of the Psalms never got the news. This is somewhat perplexing since, as Galileo showed a year before the trial, a state of absolute rest is impossible to verify experimentally. Any experiment that you perform at rest, such as throwing a ball up in the air and catching it, will have an identical result if performed while moving at a constant speed, as, say, might happen while riding on an airplane in the absence of turbulence. No experiment you can perform on the plane, if its windows are closed, will tell you whether the plane is moving or standing still.

While Galileo started the ball rolling, both literally and metaphorically, in 1632, it took another 273 years to fully lay to rest this issue (issues, unlike objects, *can* be laid to rest). It would take Albert Einstein to do so.

Einstein was not a revolutionary in the same sense as Galileo, if by this term one describes those who tear down the dictates of the authorities who came before, as Galileo had done for Aristotle. Einstein did just the opposite. He knew that rules that had been established on the basis of experiment could not easily be tossed aside, and it was a mark of his genius that he didn't.

This is so important I want to repeat it for the benefit of those people who write to me every week or so telling me that they have discovered a new theory that demonstrates everything we now think we know about the universe is wrong—and using Einstein as their exemplar to justify this possibility. Not only is your theory wrong, but you are doing Einstein a huge disservice: *rules that have been established on the basis of experiment cannot easily be tossed aside.*

· · ·

Albert Einstein was born in 1879, the same year that James Clerk Maxwell died. It is tempting to suggest that their combined brilliance was too much for one simple planet to house at the same time. But it was just a coincidence, albeit a fortuitous one. If Maxwell hadn't preceded him, Einstein couldn't have been Einstein. He came from the first generation of young scientists who grew up wrestling with the new knowledge about light and electromagnetism that Faraday and Maxwell had generated. This was the true forefront of physics for young Turks such as Einstein near the end of the nineteenth century. Light was on everyone's mind.

Even as a teenager, Einstein was astute enough to realize that Maxwell's beautiful results regarding the existence of electromagnetic waves presented a fundamental problem: they were inconsistent with the equally beautiful and well-established results of Galileo regarding the basic properties of motion, produced three centuries earlier.

Even before his epic battle with the Catholic Church over the motion of Earth, Galileo had argued that no experiment exists that can be performed by anyone to determine whether he or she is moving uniformly or standing still. But up until Galileo, a state of absolute rest was considered special. Aristotle had decided that all objects sought out the state of rest, and the Church decided that rest was so special that it should be the state of the center of the universe, namely the planet on which God had placed us.

Like a number of Aristotle's assertions, although by no means all, this notion that a state of rest is special is quite intuitive. (For those who like to quote Aristotle's wisdom when appealing to his "Prime Mover" argument for the existence of God, let us remember that he also claimed that women had a different number of teeth than men, presumably without bothering to check.)

Everything we see in our daily lives comes to rest. Everything, that is, except the Moon and the planets, which is perhaps one reason that these were felt to be special in antiquity, guided by angels or gods.

However, every sense that we have that we are at rest is an illusion. In the example I gave earlier of throwing a ball up and catching it while in a moving plane, you will eventually be able to tell that your plane is moving when you feel the bouncing of turbulence. But even when the plane is on the tarmac, it is not at rest. The airport is moving with the Earth at about 30 km/sec around the Sun, and the Sun is moving about 200 km/sec around the galaxy, and so on.

Galileo codified this with his famous assertion that the laws of physics are the same for all observers moving in a uniform state of motion, i.e., at a constant velocity in a straight line. (Observers at rest are simply a special case, when velocity is zero.) By this he meant that there is no experiment you can perform on such an object that can tell you it is not at rest. When you look up in the air at an airplane, it is easy to see that it is moving relative to you. But, there is no experiment you can perform on the ground or on the plane that will distinguish whether the ground on which you are standing is moving past the plane, or vice versa.

While it seems remarkable that it took so long for anyone to recognize this fundamental fact about the world, it does defy most of our experience. Most, but not all. Galileo used examples of balls rolling down inclined planes to demonstrate that what previous philosophers thought was fundamental about the world—the retarding force of friction that makes things eventually settle at rest—was not fundamental at all but rather masked an underlying reality. When balls roll down one plane and up another, Galileo noted, on smooth surfaces the balls would rise back to the same height at which they started. But by considering balls rolling up planes of ever-decreasing incline, he showed that the balls would have to roll farther to reach their same original height. He then reasoned that if the second incline disappeared entirely, the balls would continue rolling at the same speed forever.

This realization was profoundly important and fundamentally changed much about the way we think about the world. It is often simply called the Law of Inertia, and it set up Newton's law of motion, relating the magnitude of an external force to the observed acceleration of an object. Once Galileo recognized that it took no force to keep something moving at a constant velocity, Newton could make the natural leap to propose that it took a force to change its velocity.

The heavens and the Earth were no longer fundamentally different. The hidden reality underlying the motion of everyday objects also made clear that the unending motion of astronomical objects was not supernatural, setting the stage for Newton's Universal Law of Gravity, further demoting the need for angels or other entities to play a role in the cosmos.

Galileo's discovery was thus fundamental to establishing physics as we know it today. But so was Maxwell's later brilliant unification of electric and magnetic forces, which established the mathematical framework on which all of current theoretical physics is built.

. . .

As Albert Einstein began his journey in this rich intellectual landscape, he quickly spied a deep and irreconcilable chasm running through it: both Galileo and Maxwell could not be right at the same time.

More than twenty years ago, when my daughter was an infant, I first began to think about how to explain the paradox that young Einstein struggled with, and a good example literally hit me on the head while driving her in my car.

Galileo had demonstrated that as long as I am driving safely and at a constant speed and not accelerating suddenly, the laws of physics in our car should be indistinguishable from the laws of physics that would be measured in the laboratories in the physics building to which I was driving to work. If my daughter was playing with a toy in the backseat, she could throw the toy up in the air and expect to catch it without any surprises. The intuition her body had built up to play at home would have served her well in the car.

However, riding in the car did not lull her to sleep like many young children, but rather made her anxious and uncomfortable. During our trip, she got sick and projectile-vomited, and the vomit followed a trajectory well described by Newton, with an initial speed of, say, fifteen miles per hour, and a nice parabolic trajectory in the air, ending on the back of my head.

Say my car was coasting to a red light at this time at a relatively slow speed, say, ten miles per hour. Someone on the ground watching all of this would see the vomit traveling at 25 miles per hour, the speed of the car relative to them (10 mph) plus the speed of the vomit (15 mph), and its trajectory would be well described by Newton again, with this higher speed (25 mph) as it traveled toward my (now moving) head.

So far so good. Here's the problem, however. Now that my daughter is older, she loves to drive. Say she is driving behind a friend's car and dials him on her cell phone (hands-free, for safety) to tell him to turn right to get to the place they are both going. As she talks into the phone, electrons in the phone jiggle back and forth producing an electromagnetic wave (in the microwave band). That wave travels to the cell phone of her friend at the speed of light (actually it travels up to a satellite and then gets beamed down to her friend, but let's ignore that complication for the moment) and is received in time for him to make the correct turn.

Now, what would a person on the ground measure? Common sense would suggest that the microwave signal would travel from my daughter's car to her friend's car at a speed equal to the speed of light, as might be measured by a detector in my daughter's car (label it with the symbol c), plus the speed of the car.

But common sense is deceptive precisely because it is based on common experience. In everyday life we do not measure the time it takes light, or microwaves, to travel from one side of the room to another or from one phone to a nearby phone. If common sense applied here, that would mean someone on the ground (with a sophisticated measuring apparatus) would measure the electrons in my daughter's phone jiggling back and forth and observe the emanation of a microwave signal, which would be traveling at a speed c plus, say, ten miles per hour.

However, the great triumph of Maxwell was to show that he could calculate the speed of electromagnetic waves emanated by an oscillating charge purely by measuring the strength of electricity and magnetism. Therefore if the person on the ground observed the waves having speed c plus 10 mph, then for that person the strength of electricity and magnetism would have to be different from the values that my daughter would observe, for whom the waves were moving at a speed c.

But Galileo tells us this is impossible. If the measured strengths of electricity and magnetism differed between the two observers, then it would be possible to know who was moving and who was not, because the laws of physics—in this case electromagnetism—would take on different values for each observer.

So, either Galileo or Maxwell had to be right, but not both of them. Perhaps because Galileo had been working when physics was more primitive, most physicists came down closer to the side of Maxwell. They decided that the universe must have some absolute rest frame and that Maxwell's calculations applied in that frame only. All observers moving with respect to that frame would measure electromagnetic waves to have a different speed relative to themselves than Maxwell had calculated.

A long scientific tradition gave physical support to this idea. After all, if light was an electromagnetic disturbance, what was it a disturbance of? For thousands of years, philosophers had speculated about an "ether," some invisible background material filling all of space, and it became natural to suspect that electromagnetic waves were traveling in this medium, just as sound waves travel in water or air. Electromagnetic waves would travel with some fixed, characteristic speed in this medium (the speed calculated by Maxwell), and observers moving with respect to this background would observe the waves as faster or slower, depending on their relative motion.

While intuitively sensible, this notion was a cop-out, because if you think back to Maxwell's analysis, it would mean that these different observers in relative motion would measure the strength of electricity and magnetism to be different. Perhaps it was deemed to be acceptable because all speeds obtainable at the time were so small compared to the speed of light that any such differences would have been minute at best and would certainly have escaped detection.

The actor Alan Alda once turned the tables on conventional wisdom at a public event I attended by saying that art requires hard work, and science requires creativity. While both require both, what I like about his version is that it stresses the creative, artistic side of science. I would add to this statement that both endeavors require intellectual bravery.

Creativity alone amounts to nothing if it is not implemented. Novel ideas generally stagnate and die without the courage to implement them.

I bring this up here because perhaps the true mark of Einstein's genius was not his mathematical prowess (although, contrary to conventional wisdom, he was mathematically talented), but his creativity and his intellectual confidence, which fueled his persistence.

The challenge that faced Einstein was how to accommodate two contradictory ideas. Throwing one out is the easy way. Figuring out a way to remove the contradiction required creativity.

Einstein's solution was not complex, but that does not mean it was easy. I am reminded of an apocryphal story about Christopher Columbus, who got a free drink in a bar before departing to find the New World by claiming he could balance an egg upright on top of the bar. After the barman accepted the bet, Columbus broke the tip off the egg and placed it easily upright on the counter. He never mentioned not cracking it, after all.

Einstein's resolution of the Galileo-Maxwell paradox was not that different. Because, if both Maxwell and Galileo were right, then something else had to be broken to fix the picture.

But what could it be? For both Maxwell and Galileo to be right required something that was clearly crazy: in the example I gave, both observers would have to measure the velocity of the microwave emitted by my daughter's cell phone to be the same relative to them, instead of measuring values differing by the speed of the car.

However, Einstein asked himself an interesting question, What does it mean to measure the velocity of light, after all? Velocity is determined by measuring the distance something travels in a certain time. So Einstein reasoned as follows: it is possible for two observers to measure the same speed for the microwave relative to each of them, as long as the distance each measures the ray to travel relative to themselves during a fixed time interval (e.g., say, one second, as measured by each of them in their own frame of reference) is the same.

But this too is a little crazy. Consider the simpler example of the projectile vomit. Remember that in my frame it travels from her mouth in the backseat to hit my head, say, three feet away, in about one-quarter second. But for someone on the ground the car is traveling at 10 miles per hour during this period, which is about 14.5 feet per second. Thus for the person on the ground, in one-quarter second the vomit travels about 3.6 feet plus 3 feet, or a total 6.6 feet.

Hence for the two observers, the distances traveled by the vomit in the same time is noticeably different. How could it be that for the microwave the distances both observers measure could be the same?

The first hint that perhaps such craziness is possible is that electromagnetic waves travel so fast that in the time it takes the microwaves to get from one car to another, each car has moved hardly at all. Thus any possible difference in measured distance traveled during this time for the two observers would be essentially imperceptible.

But Einstein turned this argument around. He realized that both observers had not actually measured the distances traveled by the microwaves over human-scale distances, because the relevant times appropriate for light to travel over human-scale distances were so short that no one could have measured them at the time. And similarly, on human timescales light would travel such large distances that no one could measure those distances directly either. Thus, who was to say that such crazy behavior couldn't really happen?

The question then became, What is required for it to actually occur? Einstein reasoned that for this seemingly impossible result to be possible, the two different observers must measure distances and/or times differently from each other in just such a way that light, at least, would traverse the same measured distance in the same measured time for both observers. Thus, for example, it would be as if the observer on the ground in the vomit case were to measure the vomit traversing 6.6 feet, but would somehow also infer the time interval over which this happened to be larger than I would measure it inside my car, so that the inferred speed of the vomit would be the same relative to him as I measure it to be relative to me.

Einstein then made the bold assertion that something like this *does* happen, that both Maxwell and Galileo were correct, and that all observers, regardless of their relative state of motion, would measure any light ray to travel at the same speed, c, relative to them.

Of course, Einstein was a scientist, not a prophet, so he didn't just claim something outlandish on the basis of authority. He explored the consequences of his claim and made predictions that could be tested to verify it.

In doing so he moved the playing field of our story from the domain of light to the domain of intimate human experience. He not only forever changed the meaning of space and time, but also the very events that govern our lives.

Chapter 5

A STITCH IN TIME

He stretcheth out the north over the empty place,
and hangeth the earth upon nothing.

—JOB 26:7

The great epic stories of ancient Greece and Rome revolve around heroes such as Odysseus and Aeneas, who challenged the gods and often outwitted them. Things have not changed that much for more modern epic heroes.

Einstein overcame thousands of years of misplaced human perception by showing that even the God of Spinoza could not impose his absolute will on space and time, and that each of us evades those imaginary shackles every time we look around us and view new wonders amid the stars above. Einstein emulated artistic geniuses such as Vincent van Gogh and reasoned with the parsimony of Ernest Hemingway.

Van Gogh died fifteen years before Einstein developed his ideas on space and time, but his paintings make it clear that our perceptions of the world are subjective. Picasso may have had the chutzpah to claim that he painted what he saw, even as he produced representations of disjointed people with body parts pointing in different directions, but van Gogh's masterpieces demonstrate that the world can look very different to different people.

So too, Einstein explicitly argued, for the first time as far as I know in the history of physics, that "here" and "now" are observer-dependent concepts and not universal ones.

His argument was simple, based on the equally simple fact that we cannot be in two places at once.

We are accustomed to feeling that we share the same reality with those around us because we appear to share the same experiences as we look about together. But that is an illusion created by the fast speed of light.

When I observe something happening now, say, a car crash down the street or two lovers kissing under a lamppost as I walk nearby, neither of these events happened now, but rather then. The light that enters my eye was reflected off the car or the people just a little bit earlier.

Similarly when I take a photo of a beautiful landscape, as I just did in Northern Ireland where I began writing this chapter, the scene I captured is not a scene merely spread out in space, but rather in space *and* time. The light from the distant pillared cliffs at Giant's Causeway perhaps a kilometer away left those cliffs well before (perhaps thirty-millionths of a second before) the light from the people in the foreground scrambling over the hexagonal lava pods left to reach my camera at the same time.

With this realization, Einstein asked himself what two events that one observer views as happening at the same time in two different locations would look like for another observer moving with respect to the first observer while the observations were being made. The example he considered involved a train, because he lived in Switzerland at a time when a train was leaving about every five minutes for somewhere in the country from virtually any other place in the country.

Imagine the picture shown below in which lightning hits two points beside either end of a train that are equidistant from observer A, who is at rest with respect to those points, and observer B on a moving train, who passes by A at the instant A later determines the lightning bolts struck:

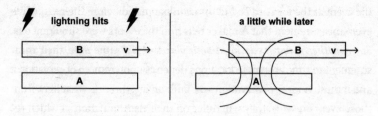

A little while later A will see both lightning flashes reaching him at the same time. B, however, will have moved during this time. Therefore the light wave bringing the information that a flash occurred on the right will already have passed B, and the light bringing the information about the flash on the left will not yet have reached him.

B sees the light coming from either end of his train, and indeed the flash at the front end occurs before the flash at the rear end. Since he measures the light as traveling toward him at speed c, and since he is in the middle of his train, he concludes therefore that the right-hand flash must have occurred before the left-hand flash.

Who is right here? Einstein had the temerity to suggest that both observers were right. If the speed of light were like other speeds, then B would of course see one wave before the other, but he would see them traveling toward him at different speeds (the one he was moving toward would be faster and the one from which he was moving away would be slower), and he would therefore infer that the events happened at the same time. But because both light rays are measured by B to be traveling toward him at the same speed, c, the reality he infers is completely different.

As Einstein pointed out, when defining what we mean by different physical quantities, measurement is everything. Imagining a reality that is independent of measurement might be an interesting philosophical exercise, but from a scientific perspective it is a sterile line of inquiry. If both A and B are located at the same place at the same time, they must both measure the same thing at that instant, but if they are in remote locations, almost all bets are off. Every measurement that B can make tells him that the event at the forward end of his train happened before the next, while every measurement that A makes tells him the events were simultaneous. *Since neither A nor B can be at both places at the same time*, their measurement of time at remote locations depends upon remote observations, and if those remote observations are built on interpreting what light from those events reveals, they will differ on their determination of which remote events are simultaneous, and they will both be correct.

Here and now is only universal for here and now, not there and then.

. . .

I wrote "almost all" bets are off for a reason. For as strange as the example I just gave might seem, it can actually be far stranger. Another observer, C, traveling on a train moving in the opposite direction from B on a third track beside A and B will infer that the event on the left side (the forward part of his train) occurred before the event on the right-hand side. In other words, the order of the events seen by the two observers B and C will be completely reversed. One person's "before" will be the other's "after."

This presents a big apparent problem. In the world in which most of us believe we live, causes happen *before* effects. But if "before" and "after" can be observer dependent, then what happens to cause and effect?

Remarkably, the universe has a sort of built-in catch-22, which ends up ensuring that while we need to keep an open mind about reality, we don't have to keep it so open that our brains fall out, as the publisher of the *New York Times* used to say. In this case, Einstein demonstrated that a reversal of the time ordering of distant events brought about by the constancy of light is only possible if the events are far enough apart so that a light ray will take longer to travel between them than the inferred time difference between the two events. Then, if nothing can travel faster than light (which turns out to be another consequence of Einstein's effort to coordinate Galileo and Maxwell), no signal from one event could ever arrive in time to affect the other, so one event could not be the cause of the other.

But what about two different events that occur some time apart at the same place. Will different observers disagree about them? To analyze this situation Einstein imagined an idealized clock on a train. The ticks of the clock occur each time a light ray sent from a clock on one side of the train reflects off a mirror located on the other side and returns to the clock on the original side of the train (see below).

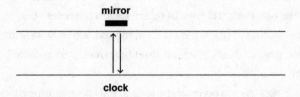

Let us say each round-trip (tick) is a millionth of a second. Now consider an observer on the ground watching the same round-trip. Because the train is moving, the light ray travels on the trajectory shown below, with the clock and mirror having moved between the time of emission and reception.

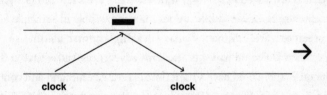

Clearly this light ray traverses a greater distance relative to the observer on the ground than it does relative to the clock on the train. However, the light ray is measured to be traveling at the same speed, c. Thus, the round-trip takes longer. As a result, the one-millionth-of-a-second click of the clock on the train is observed on the ground to take, say, two-millionths of a second. The clock on the train is therefore ticking at half the rate of a clock on the ground. Time has slowed down for the clock on the train.

Stranger still, the effect is completely reciprocal. Someone aboard the train will observe a clock on the ground as ticking at half the rate of their clock on the train, as the figure would look identical for someone on the train watching a light travel between mirrors placed on the ground.

This may make it seem like the slowing of clocks is merely an illusion, but once again, measurement equals reality, although in this case a little more subtly than for the case of simultaneity. To compare clocks later to see which, if any, of the observers' clocks has really slowed down, at least one of the observers will have to return to join the other. That observer will have to change his or her uniform motion, either by slowing down and reversing, or by speeding up from (apparent) rest and catching up with the other observer.

This makes the two observers no longer equivalent. It turns out that the observer who does the accelerating or the decelerating will find, when she arrives back at the starting position, that she has actually aged far less than her counterpart, who has been in uniform motion during the whole time.

This sounds like science fiction, and indeed it has provided the fodder for a great deal of science fiction, both good and bad, because in principle it allows for precisely the kind of space travel around the galaxy that is envisaged in so many movies. There are a few rather significant glitches, however. While it does make it possible in principle for a spacecraft to travel around the galaxy in a single human lifetime, so that Jean-Luc Picard could have his *Star Trek* adventures, those back at Star Fleet command would have a hard time exerting command and control over any sort of federation. The mission of ships such as the USS *Enterprise* might be five years long for the crew on board, but each round-trip from Earth to the center of the galaxy of a ship at near light speed would take sixty thousand years or so as experienced by society back home. To make matters worse, it would take more fuel than there is mass in the galaxy to power a single such voyage, at least using conventional rockets of the type now in use.

Nevertheless, science fiction woes aside, "time dilation"—as the relativistic slowing of clocks is called with regard to moving objects—is very much real, and very much experienced every day here on Earth.

At high-energy particle accelerators such as the Large Hadron Collider, for example, we regularly accelerate elementary particles to speeds of 99.9999 percent of the speed of light and rely on the effects of relativity when exploring what happens.

But even closer to home, relativistic time dilation has an impact. We on Earth are all bombarded every day by cosmic rays from space. If you had a Geiger counter and stood out in a field, the counter would click at a regular rate every few seconds, as it recorded the impact of high-energy particles called muons. These particles are produced where high-energy protons in cosmic rays smash into the atmosphere, producing a shower of other, lighter particles—including muons—which are unstable, with a lifetime of about one-millionth of a second, and decay into electrons (and my favorite particles, neutrinos).

If it weren't for time dilation, we would never detect these muon cosmic rays on Earth. Because a muon traveling at close to the speed of light for a millionth of a second would cover about three hundred meters before decaying. But the muons raining down on Earth make it twenty kilometers, or about twelve and a half miles or so, from the upper atmosphere, in which they are produced, down to our Geiger counter. This is possible only if the muons' internal "clocks" (which prompt them to decay after one-millionth of a second or so) are ticking slowly relative to our clocks on Earth, ten to one hundred times more slowly than they would be if they were produced at rest here in a laboratory on Earth.

. . .

The last implication of Einstein's realization that the speed of light must be constant for all observers appears even more paradoxical than the others—in part because it involves changing the physical behavior of objects we can see and touch. But it also will help carry us back to our beginnings to glimpse a new world beyond the confines of our normal earthbound imagination.

The result is simply stated, even if the consequences may take some time to digest. When I am carrying an object such as a ruler, and moving fast compared to you, my ruler will be measured by you to be smaller than it is for me. I might measure it to be 10 cm, say:

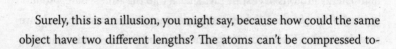

But to you, it might appear to be merely 6 cm:

Surely, this is an illusion, you might say, because how could the same object have two different lengths? The atoms can't be compressed together for you, but not for me.

Once again, we return to the question of what is "real." If every measurement you can perform on my ruler tells you it is 6 cm long, then it *is* 6 cm long. "Length" is not an abstract quantity but requires a measurement. Since measurement is observer dependent, so is length. To see this is possible while illuminating another of relativity's slippery catch-22s, consider one of my favorite examples.

Say I have a car that is twelve feet long, and you have a garage that is eight feet deep. My car will clearly not fit in your garage:

But, relativity implies that if I am driving fast, you will measure my car to be only, say, six feet long, and so it should fit in your garage, at least while the car is moving:

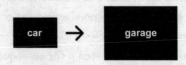

However, let's view this from my vantage point. For me, my car is twelve feet long, and your garage is moving toward me fast, and it now is measured by me to be not eight feet deep, but rather four feet deep:

Thus, my car clearly cannot fit in your garage.

So which is true? Clearly my car cannot both be inside the garage and not inside the garage. Or can it?

Let's first consider your vantage point, and imagine that you have fixed big doors on the front of your garage and the back of your garage. So that I don't get killed while driving into it, you perform the following. You have the back door closed but open the front door so my car can drive in. When it is inside, you close the front door:

However, you then quickly open the back door before the front of my car crashes, letting me safely drive out the back:

Thus, you have demonstrated that my car was inside your garage, which of course it was, because it is small enough to fit in it.

However, remember that, for me, the time ordering of distant events can be different. Here is what I will observe.

I will see your tiny garage heading toward me, and I will see you open the front door of the garage in time for the front of my car to pass through.

I will then see you kindly open the back door before I crash:

After that, and after the back of my car is inside the garage, I will see you close the front door of your garage:

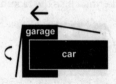

As will be clear to me, my car was never inside your garage with both doors closed at the same time because that is impossible. Your garage is too small.

"Reality" for each of us is simply based on what we can measure. In my frame the car *is* bigger than the garage. In your frame the garage *is* bigger than my car. Period. The point is that we can only be in one place at one time, and reality where we are is unambiguous. But what we infer about the real world in other places is based on remote measurements, which are observer dependent.

But the virtue of careful measurement does not stop there.

The new reality that Einstein unveiled, based as it was on the empirical validity of Galileo's law, and Maxwell's remarkable unification of electricity and magnetism, appears on its face to replace any last vestige of objective reality with subjective measurement. As Plato reminds us, however, the job of the natural philosopher is to probe deeper than this.

It is said that fortune favors the prepared mind. In some sense, Plato's cave prepared our minds for Einstein's relativity, though it remained for Einstein's former mathematics professor Hermann Minkowski to complete the task.

Minkowski was a brilliant mathematician, eventually holding a chair at the University of Göttingen. But in Zurich, where he was one of Einstein's professors, he was a brilliant mathematician whose classes Einstein skipped, because while he was a student, Einstein appeared to have a great disdain for the significance of pure mathematics. Time would change that view.

Recall that the prisoners in Plato's cave also saw from shadows on their wall that length apparently had no objective constancy. The shadow of a ruler might at one time look like this, at 10 cm:

and, at another time like this, at 6 cm:

The similarity with the example I presented when discussing relativity is intentional. In the case of Plato's cave dwellers, however, we recognized that this length contraction occurred because the cave dwellers were merely seeing two-dimensional shadows of an underlying three-dimensional object. Viewed from above, it can easily be seen that the shorter shadow projected on the wall results because the ruler has been rotated at an angle to the wall:

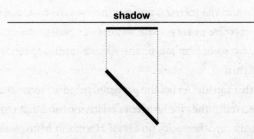

And as another Greek philosopher, Pythagoras, taught us, when seen this way, the length of the ruler is fixed, but the projections onto the wall and a line perpendicular to the wall always combine together to give the same length, as shown below:

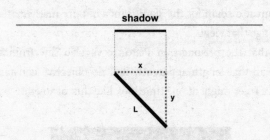

This yields the famous Pythagorean theorem, $L^2 = x^2 + y^2$, which high school students have been subjected to for as long as high schools have taught geometry. In three dimensions, this becomes $L^2 = x^2 + y^2 + z^2$.

Two years after Einstein wrote his first paper on relativity, Minkowski recognized that perhaps the unexpected implications of the constancy of the speed of light, and the new relations between space and time unveiled by Einstein, might also reflect a deeper connection between the two. Knowing that a photograph, which we usually picture as a two-dimensional representation of three-dimensional space, is really an image spread out in both space and time, Minkowski reasoned that observers who were moving relative to each other might be observing *different* three-dimensional slices of a four-dimensional universe, one in which space and time are treated on an equal footing.

If we return to the ruler example in the case of relativity, where the ruler of the moving observer is measured to be shorter by the other observer than it would be in the frame in which it is at rest, we should also remember that for this observer the ruler is also "spread out" in time—events at either end that are simultaneous to the observer at rest with respect to the ruler are not simultaneous for the second observer.

Minkowski recognized that one could accommodate this fact, and all the others, by considering that the different three-dimensional perspectives probed by each observer were in some sense different "rotated" projections of a four-dimensional "space-time," where there exists an invariant four-dimensional space-time "length" that would be the same for all observers. The four-dimensional space, which we now call Minkowski space, is a little different from its 3-D counterpart, in that time as a fourth dimension is treated slightly differently from the three dimensions of space, x, y, and z. The four-dimensional "space-time length," which we can label as S, is *not* written, in analogy to the three-dimensional length, which we denoted by L, above, as

$$S^2 = x^2 + y^2 + z^2 + t^2$$

but rather as

$$S^2 = x^2 + y^2 + z^2 - t^2.$$

The minus sign that appears in front of t^2 in the definition of space-time length, S, gives Minkowski space its special characteristics, and it is the reason our different perspectives of space and time when we are moving relative to one another are not simple rotations, as in the case of Plato's cave, but something a little more complicated.

Nevertheless, in one fell swoop, the very nature of our universe had changed. As Minkowski poetically put it in 1908: "Henceforth space by itself, and time by itself, are doomed to fade away into mere shadows, and only a kind of union of the two will preserve an independent reality."

Thus, on the surface, Einstein's Special Theory of Relativity appears to make physical reality subjective and observer dependent, but *relativity* is in this sense a misnomer. The Theory of Relativity is instead a theory of absolutes. Space and time measurements may be subjective, but "space-time" measurements are universal and absolute. The speed of light is universal and absolute. And four-dimensional Minkowski space is the field on which the game of nature is played.

The depth of the radical change in perspective brought about by Minkowski's reframing of Einstein's theory can perhaps best be understood by considering Einstein's own reactions to Minkowski's picture. Initially Einstein called it "superfluous learnedness," suggesting that it was simply fancy mathematics, devoid of physical significance. Shortly thereafter he emphasized this by saying, "Since the mathematicians have invaded relativity theory, I do not understand it myself anymore." Ultimately, however, as happened several times in his lifetime, Einstein came around and recognized that this insight was essential to understand the true nature of space and time, and he later built his General Theory of Relativity on the foundation that Minkowski had laid.

It would have been difficult if not impossible to guess that Faraday's spinning wheels and magnets would eventually lead to such a profound revision in our understanding of space and time. With the spectacles of hindsight, however, we could have had at least an inkling that the unification of electricity and magnetism could have heralded a world where motion would reveal a new underlying reality.

Returning to Faraday and Maxwell, one of the important discoveries that started the ball rolling was that a magnet acts on a moving electric charge with an odd force. Instead of pushing the charge forward or backward, the magnet exerts a force always at right angles to the motion of the electric charge. This force, now called the Lorentz force—after

Hendrik Lorentz, a physicist who came close to discovering relativity himself—can be pictured as follows:

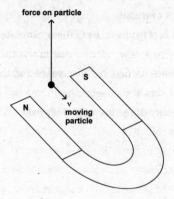

The charge moving between the poles of the magnet gets pushed upward.

But now consider how things would look from the frame of the particle. In its frame, the magnet would be moving past it.

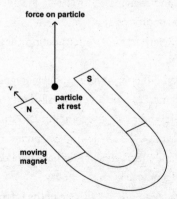

But by convention we think of an electrically charged particle at rest as being affected only by electric forces. Thus, since the particle is at rest in this frame, the force pushing the particle upward in this picture would be interpreted as an electric force.

One person's magnetism is therefore another person's electricity, and what connects the two is motion. The unification of electricity and magnetism reflects at its heart that uniform relative motion gives observers different perspectives of reality.

Motion, a subject first explored by Galileo, ultimately provided, three centuries later, a key to a new reality—one in which not only electricity and magnetism were unified, but also space and time. No one could have anticipated this saga at its beginning.

But that is the beauty of the greatest story ever told.

Chapter 6

THE SHADOWS OF REALITY

As they were walking along and talking together,
suddenly a chariot of fire and horses of fire
appeared and separated the two of them.

—2 KINGS 2:11

One might have thought that, in 1908, following the after-shock of the discovery of an unexpected hidden connection between space and time, nature couldn't have gotten much stranger. But the cosmos doesn't care about our sensibilities. And once again, light provided the key to the door of the rabbit hole to a world that makes Alice's experiences seem tame.

While they may be strange, the connections unearthed by Einstein and Minkowski can be intuitively understood—given the constancy of the speed of light—as I have tried to demonstrate. Far less intuitive was the next discovery, which was that on very small scales, nature behaves in a way that human intuition cannot ever fully embrace, because we cannot directly experience the behavior itself. As Richard Feynman once argued, no one understands quantum mechanics—if by *understand* one means developing a concrete physical picture that appears fully intuitive.

Even many years after the rules of quantum mechanics were discovered, the discipline would keep yielding surprises. For example, in 1952 the astrophysicist Hanbury Brown built an apparatus to measure the angular size of large radio sources in the sky. It worked so well that he and a colleague, Richard Twiss, applied the same idea to try to measure the optical light from individual stars and determine their angular size. Many physicists claimed that their instrument, called an intensity interferometer, could not possibly work. Quantum mechanics, they argued, would rule it out.

But it worked. It wasn't the first time physicists had been wrong about quantum mechanics, and it wouldn't be the last. . . .

Coming to grips with the strange behavior of quantum mechanics means often accepting the seemingly impossible. As Brown himself amusingly put it when trying to explain the theory of his intensity interferometer, he and Twiss were expounding the "paradoxical nature of light, or if you like, explaining the incomprehensible—an activity closely, and interestingly, analogous to preaching the Athanasian Creed." Indeed, like many of the stranger effects in quantum mechanics, the Holy Trinity—Father, Son, and Holy Ghost all embodied at the same time in a single being—is also seemingly impossible. The similarity ends there, however.

Common sense also tells us that light cannot be both a wave and a particle at the same time. However, in spite of what common sense suggests, and whether we like it or not, experiments tell us it is so. Unlike the Creed, developed in the fifth century, this fact is not a matter of semantics or choice or belief. So we don't need to recite quantum mechanics creeds every week to make them seem less bizarre or more believable.

One hears about the "interpretation of quantum mechanics" for good reason, because the "classical" picture of reality—namely the picture given by Newton's laws of classical motion of the world as we experience it on human scales—is inadequate to capture the full picture. The surface world we experience hides key aspects of the processes that underlie the phenomena we observe. So too Plato's philosophers could not discover the biological processes that govern humans by observing just the shadows of humans on the wall. No level of analysis would be likely to allow them to intuit the full reality underlying the dark forms.

The quantum world defies our notion of what is sensible—or even possible. It implies that at small scales and for short times, the simple classical behavior of macroscopic objects—baseballs thrown from pitcher to catcher, for example—simply breaks down. Instead, on small scales, objects are undergoing many different classical behaviors—as well as classically forbidden behaviors—at the same time.

Quantum mechanics, like almost all of physics since Plato, began with scientists thinking about light. So it is appropriate to begin to explore quantum craziness by starting with light, in this case by returning to an important experiment first reported by the British polymath Thomas Young around 1800—the famous "double-slit experiment."

Young lived in an era that is hard to appreciate today, when a brilliant and hardworking individual could make breakthroughs in a host of different fields. But Young was not just any brilliant hardworking individual. He was a prodigy, reading at two, and by the age of thirteen he had read the major Greek and Latin epic poems, had built a microscope and a telescope, and was learning four different languages. Later, trained as a medical doctor, Young was the first to propose, in 1806, the modern concept of energy, which now permeates every field of scientific endeavor. That alone would have made him memorable, but in his spare time he also was one of the first to help decipher the hieroglyphics on the Rosetta stone. He developed the physics of elastic materials, associated with what is now called Young's modulus, and helped first elucidate the physiology of color vision. And his brave demonstration of the wave nature of light (which argued against Isaac Newton's powerful claim that light was made of particles) was so compelling that it helped lay the basis of Maxwell's discovery of electromagnetic waves.

Young's experiment is simple. Let's return to Plato's cave and consider a screen placed in front of the back wall of the cave. Place two slits in the screen as shown below (as seen from above):

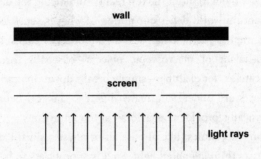

If the light is made of particles, then those light rays that pierce the slits would form two bright lines on the wall behind these two slits:

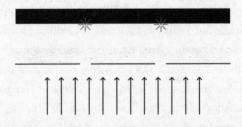

However, it was well known that waves, unlike particles, diffract around barriers and narrow slits and would produce a very different pattern on the wall. If waves impinge on the barrier, and if each slit is narrow, a circular pattern of waves is generated at each slit, and the patterns from the two slits can "interfere" with each other, sometimes constructively and sometimes destructively. The result is a pattern of bright and dark regions on the back wall, as shown below:

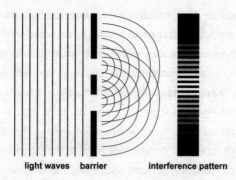

light waves **barrier**   **interference pattern**

Using just such an apparatus, with narrow slits, Young reported this interference pattern, characteristic of waves, and so definitively demonstrated the wave nature of light. In 1804, this was a milestone in the history of physics.

One can try the same experiment that Young tried for light on elementary particles such as electrons. If we send a beam of electrons toward a phosphorescent screen, like the screen in old-fashioned television sets, you will see a bright dot where the beam hits the screen. Now imagine that we put two slits in front of the screen, as Young did for light, and aim a wide stream of electrons at the screen:

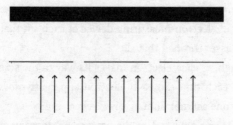

Here, based on the reasoning I gave when I discussed the behavior of light, you would expect to see a bright line behind each of the two slits, where the electrons could pass through to the screen. However, as you have probably already guessed, this is not what you would see, at least if the slits are narrow enough and close enough. Instead, you see an interference pattern similar to that which Young observed for light waves. Electrons, which are particles, seem to behave in this case just like waves of light. In quantum mechanics, particles have wavelike properties.

That the electron "waves" emanating from one slit can interfere with electron "waves" emanating from the other slit is unexpected and strange, but not nearly as strange as what happens if we send a stream of electrons toward the screen *one at a time*. Even in this case, the pattern that builds up on the screen is identical to the interference pattern. Somehow, each electron interferes with itself. Electrons are not billiard balls.

We can understand this as follows: The probability of an electron's hitting the screen at each point is determined by treating each electron as not taking a single trajectory, but rather following many different trajectories *at once*, some of which go through one slit and some of which go through the other. Those that go through one slit then interfere with those that go through the other slit—producing the observed interference pattern at the screen.

Put more bluntly, one cannot say the electron goes through either one slit or the other, as a billiard ball would. Rather it goes through neither and at the same time it goes through both.

Nonsense, you insist. So you propose a variant of the experiment to prove it. Put an electron-measuring device at each slit that clicks when an electron passes through that slit.

Sure enough, as each electron makes its way to the screen, only one device clicks each time. So each electron apparently *does* go through one and only one slit, not both.

However, if you now *look* at the pattern of electrons accumulating at the screen behind the slits, the pattern will have changed from the original interference pattern to the originally expected pattern—with a bright region behind each of the two slits, just as if one were shooting billiard balls or bullets and not waves toward the screen.

In other words, in attempting to verify your classical intuition, you changed the behavior of the electrons. Or, as more commonly asserted in quantum mechanics, measurement of a system can alter its behavior.

One of the many seemingly impossible aspects of quantum mechanics is that there is no experiment you can perform that demonstrates that in the absence of measurement the electrons behave in a sensible classical way.

This strange wavelike nature of objects that would otherwise be considered to be particles—such as electrons—is mathematically expressed by assigning to each electron a "wave function," which describes the probability of finding that electron at any given point. If the wave function takes on non-zero values at many different points, then the electron's position cannot be isolated *in advance of accurately measuring its position*. In other words there is a non-zero probability that the electron is not actually localized at just some specific point in space in advance of making a measurement.

While you might imagine that this is a simple problem of not having access to all the information we need to locate the particle until we make a measurement, Young's double-slit experiment, when updated for electrons, demonstrated that this is most certainly not the case. Any "sensible" classical picture of what is happening between measurements is inconsistent with the data.

. . .

The strange behavior of electrons was not the first evidence that the microscopic world could not be understood by intuitive classical logic. Once again, in keeping with the revolutionary developments in our understanding of nature since Plato, the discovery of quantum mechanics began with a consideration of light.

Recall that if we perform Young's double-slit experiment in Plato's cave with light rays, we get the interference pattern on the wall that Young discovered, which demonstrated that light was indeed a wave. So far, so good. However, if the light source is sufficiently weak, then if we try to detect the light as it passes through either of the slits, something strange happens. We will measure the light beam as traveling through one slit or the other, not both. And as with electrons, in this case the pattern on the wall will now change, looking as it would if light were particles and not waves.

In fact, light also behaves like both a particle and a wave, depending on the circumstances under which you choose to measure it. The individual particles of light, which we now call photons, were first labeled *quanta* by the German theoretical physicist Max Planck, who suggested in 1900 that light might be emitted or absorbed in some smallest bundle (although the idea that light might come in discrete packets had earlier been floated by the great Ludwig Boltzmann in 1877).

I have come to admire Planck even more as I have learned about his life. Like Einstein, he was an unpaid lecturer and was not offered an academic position after completing his thesis. During this time he spent his career trying to understand the nature of heat and developed several important pieces of work in thermodynamics. Five years after defending his thesis, he was finally offered a university position, and he then quickly rose up the ranks and became a full professor at the prestigious University of Berlin in 1892.

In 1894 he turned to the question of the nature of light emitted by hot objects, in part driven by commercial considerations (the first example I know of in the story I have been telling where fundamental physics was commercially motivated). He was commissioned to explore how to get the maximum amount of light out of the newly invented lightbulbs while using the minimum amount of energy.

We all know that when we heat up an oven element it first glows red, and then, when it gets hotter, it begins to glow blue. But why? Surprisingly, the conventional approaches to this problem were unable to reproduce these observations. After struggling with the problem for six years, Planck presented a revolutionary proposal about radiation that agreed with observations.

Originally there was nothing revolutionary about his derivation, but within two months he had revised his analysis to accommodate ideas about what was happening at a fundamental level. In a quote that has endeared him to me since I first read it, he wrote that his new approach arose as "an act of despair. . . . I was ready to sacrifice any of my previous convictions about physics."

This reflects to me the fundamental quality that makes the scientific process so effective, and which is so clearly represented in the rise of quantum mechanics. "Previous convictions" are just convictions waiting to be overturned—by empirical data, if necessary. We throw out cherished old notions like yesterday's newspaper if they don't work. And they didn't work in explaining the nature of radiation emitted by matter.

Planck derived his law of radiation from the fundamental assumption that light, which was a wave, nevertheless was emitted only in "packets" of some minimum energy—proportional to the frequency of the radiation in question. He labeled the constant that related the energy to the frequency the "action quantum," which is now called Planck's constant.

This may not sound so revolutionary, and as Faraday did with electric fields, Planck viewed his assumption as merely a formal mathematical crutch to aid in his analysis. He later stated, "Actually I did not think much about it." Nevertheless, this proposal that light was emitted in particle-like packets is clearly difficult to reconcile with the classical picture of light as a wave. The energy carried by a wave is simply related to the magnitude of its oscillations, which can change continuously from zero. However, according to Planck, the amount of energy that could be emitted in a light wave of a given frequency had an absolute minimum. This minimum was termed an "energy quantum."

Planck subsequently tried to develop a classical physical understanding of these energy quanta, but failed—causing him, as he put it, "much trouble." Still, unlike a number of his colleagues, he recognized that the universe didn't exist to make his life easier. Referring to the physicist and astronomer Sir James Jeans, who was unwilling to give up classical notions in the face of the evidence provided by radiation, Planck stated, "I am unable to understand Jeans's stubbornness—he is an example of a theoretician as should never be existing, the same as Hegel was for philosophy. So much the worse for the facts if they don't fit." (Just to be clear, in case readers are moved to write me letters, Planck cast this aspersion on Hegel, not me!)

Planck later became friends with another physicist who had let the facts drive him toward another revolutionary idea, Albert Einstein. In 1914, when Planck had become dean at Berlin University, he established a new professorship for Einstein there. At first Planck could not accept Einstein's remarkable proposal—made in 1905, the same year in which he proposed the Special Theory of Relativity—that not only was light emitted by matter in quantum packets, but that light beams themselves existed as bunches of these quanta—that light itself was made up of particle-like objects, which we now call photons.

Einstein was driven to this proposal to explain a phenomenon called the photoelectric effect, discovered by Philipp Lenard in 1902—a physicist whose anti-Semitism would later play a key role in delaying Einstein's Nobel Prize, and ensuring, curiously, if perhaps poetically, that it would be not for Einstein's work on relativity, but rather on the photoelectric effect. In the photoelectric effect, light shining on a metal surface can knock electrons out of atoms and produce a current. However, no matter how intense the light, no electrons would be emitted if the frequency of the light was below some threshold. The moment the frequency was raised above that threshold, a photoelectric current would be generated.

Einstein realized, correctly, that this could be explained if the light came in minimum packets of energy, with the energy proportional to the frequency of light—as Planck had postulated for light emitted by matter. In this case, only light with frequencies greater than some threshold frequency could contain quanta energetic enough to kick electrons out of atoms.

Planck could accept the quantized emission of radiation as explaining his radiation law, but the assumption that light itself was quantum-like (i.e., particle-like) was so foreign to the common understanding of light as an electromagnetic wave that Planck balked. Only six years later, at a conference in Belgium, the Solvay Conference, which later became famous, was Einstein finally able to convince Planck that the classical picture of light had to be abandoned, and that quanta—aka photons—were real.

Einstein was also the first to actually use a fact that he later denounced in his famous statement deriding the probabilistic essence of quantum mechanics and reality: "God does not play dice with the universe." He showed that if atoms *spontaneously* (i.e., *without direct cause*) absorb and emit finite packets of radiation as electrons jump between discrete energy levels in atoms, then he could rederive the Planck radiation law.

It is ironic that Einstein, who started the quantum revolution but never joined it, was also perhaps the first to use probabilistic arguments to describe the nature of matter—a strategy that the subsequent physicists who turned quantum mechanics into a full theory would place front and center. As a result, Einstein was one of the first physicists to demonstrate that God *does* play dice with the universe.

To take the analogy a little further, Einstein was one of the first physicists to demonstrate that the classical notion of causation begins to break down in the quantum realm. Many people have taken exception to my proposal that the universe needed no cause but simply popped into existence from nothing. Yet this is precisely what happens with the light you are using to read this page. Electrons in hot atoms emit photons—photons that didn't exist before they were emitted—which are emitted spontaneously and without specific cause. Why is it that we have grown at least somewhat comfortable with the idea that photons can be created from nothing without cause, but not whole universes?

The realization that electromagnetic waves were also particles began a quantum revolution that would change everything about the way we view nature. To be a particle and a wave at the same time is impossible classically—as should be clear from the earlier discussion in this chapter—but it is possible in the quantum world. As should also be clear, this was just the beginning.

Chapter 7

A UNIVERSE STRANGER
THAN FICTION

Therefore do not throw away your confidence,
which has a great reward.

—HEBREWS 10:35

Conventional wisdom might suggest that physicists love
to invent crazy esoterica to explain the universe around us, either because we have nothing better to do, or because we are particularly per-
verse. However, as the unveiling of the quantum world demonstrates,
more often than not it is nature that drags us scientists, kicking and
screaming, away from the safety of what is familiar.

Nevertheless, to say that the pioneers who pushed us forward into the
quantum world lacked confidence would be a profound misstatement. The
voyage they embarked upon was without precedent and without guides.
The world they were entering defied all common sense, and classical logic,
and they had to be prepared at every turn for a change in the rules.

Imagine taking a road trip to another country, where the inhabitants
all speak a foreign language, and the laws are not based on experiences
that compare to any you have ever had in your life. Moreover imagine
the traffic signals are hidden and can change from place to place. Then

you can get a sense of where the young Turks who overturned our understanding of nature in the first half of the twentieth century were heading.

The analogy between exploring strange new quantum worlds and
embarking on a trek through a new landscape may seemed strained, but
exactly such a relationship between the two was paralleled in the life of
none other than Werner Heisenberg, one of the founders of quantum
mechanics, who once reminisced about an evening in the summer of
1925 on the island of Helgoland, a lovely oasis in the North Sea, when he
realized he had discovered the theory:

> *It was almost three o'clock in the morning before the final result of
> my computations lay before me. The energy principle had held for
> all the terms, and I could no longer doubt the mathematical
> consistency and coherence of the kind of quantum mechanics to
> which my calculations pointed. At first, I was deeply alarmed. I
> had the feeling that, through the surface of atomic phenomena, I
> was looking at a strangely beautiful interior and felt almost giddy
> at the thought that I now had to probe this wealth of mathematical
> structures nature had so generously spread out before me. I was far
> too excited to sleep, and so, as a new day dawned, I made for the
> southern tip of the island, where I had been longing to climb a rock
> jutting out into the sea. I now did so without too much trouble and
> waited for the sun to rise.*

Heisenberg, fresh from obtaining his PhD, had moved to the distinguished German university in Göttingen to work with Max Born to try
to come up with a consistent theory of quantum mechanics (a term first
used in the paper "On Quantum Mechanics" by Born in 1924). However,
spring hay fever had laid Heisenberg low, and he escaped the green coun-
tryside for the sea. There, he polished off his ideas about the quantum
behavior of atoms and sent it off to Born, who submitted it for publication.

You may be familiar with Heisenberg's name, not least because of the famous principle associated with it. The Heisenberg uncertainty principle has gained a New Age aura, providing fuel for many a charlatan to take advantage of people for whom quantum mechanics seems to offer hope of a world where any dream, no matter how outlandish, is realizable.

Other familiar names, Bohr, Schrödinger, Dirac, and later Feynman and Dyson, each made great leaps into the unknown. But they weren't alone. Physics is a collaborative discipline. Too often science stories are written as if the protagonists had a sudden Aha! experience alone late at night. Heisenberg had been working on quantum mechanics for several years with his PhD supervisor, the brilliant German scientist Arnold Sommerfeld (whose students would win four Nobel Prizes, and whose postdoctoral research assistants would win three), and later with Born (who was finally recognized with a Nobel almost thirty years later), as well as a young colleague, Pascual Jordan. Every major triumph we celebrate with a name and a prize is accompanied by a legion of hardworking, often less heralded, individuals, each of whom moves forward the line of scrimmage by a little bit. Baby steps are the norm, not the exception.

The most remarkable leaps into the unknown are often not fully appreciated, even by their developers, until much later. Thus Einstein, for example, never trusted his beautiful General Relativity enough to believe its prediction that the universe cannot be static but must be expanding or contracting—until observations demonstrated the expansion. And the world didn't stand on its head when Heisenberg's paper appeared. Heisenberg's friend and contemporary the brilliant and irascible physicist Wolfgang Pauli (another future Nobel laureate assistant to Sommerfeld) thought the work to be essentially mathematical masturbation, leading Heisenberg to respond in jocular form:

You have to allow that, in any case, we are not seeking to ruin physics out of malicious intent. When you reproach us that we are such big donkeys that we have never produced anything new in physics, it may well be true. But then, you are also an equally big jackass because you have not accomplished it either. . . . Do not think badly of me and many greetings.

Physics doesn't proceed in the linear fashion that textbooks recount. In real life, as in many good mystery stories, there are false leads, misperceptions, and wrong turns at every step. The story of the development of quantum mechanics is full of them. But I want to cut to the chase here, and so I will skip over Niels Bohr, whose ideas laid out the first fundamental atomic rules of the quantum world as well as the basis for much of modern chemistry. We'll also skip Erwin Schrödinger, who was a remarkably colorful character, fathering at least three children with various mistresses, and whose wave equation is the most famous icon of quantum mechanics.

Instead I will focus first on Heisenberg, or rather not Heisenberg himself, but instead the result that made his name famous: the Heisenberg uncertainty principle. This is often interpreted to mean that the observations of quantum systems affect their properties—which was manifest in our earlier discussions of electrons or photons passing through two slits and impinging on a screen behind them.

Unfortunately this leads to the misimpression that somehow observers, in particular human observers, play a key role in quantum mechanics—a confusion that has been exploited by my Twitter combatant Deepak Chopra, who, in his various ramblings, somehow seems to think the universe wouldn't exist if our consciousness weren't here to measure and frame its properties. Happily the universe predates Chopra's consciousness and was proceeding pretty nicely before the advent of all life on Earth.

However, the Heisenberg uncertainty principle at its heart has nothing to do with observers at all, even though it does limit their ability to perform measurements. It is instead a fundamental property of quantum systems, and it can be derived relatively straightforwardly and mathematically, based on the wave properties of these systems.

Consider for example a simple wavelike disturbance with a single frequency (wavelength) oscillating as it moves along the x direction:

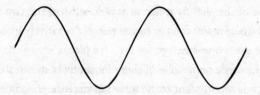

As I have noted, in quantum mechanics particles have a wavelike character. Thanks to Max Born we recognize that the square of the amplitude of the wave associated with a particle at any point—what we now call the wave function of the particle, following Schrödinger—determines the probability of finding the particle at that point. Because the amplitude of the oscillating wave above is more or less constant at all the peaks, such a wave, if it corresponded to the probability amplitude of finding an electron, would imply a more or less uniform probability for finding the electron anywhere along the path.

Now consider what a disturbance would look like if it was the sum of two waves of slightly different frequencies (wavelengths), moving along the x axis:

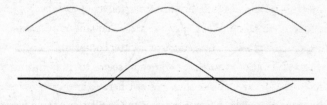

When we combine the two waves, the resulting disturbance will look like:

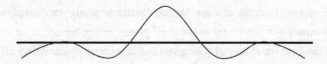

Because of the slightly different wavelengths of the two waves, the peaks and troughs will tend to cancel out, or "negatively interfere" with each other everywhere except for the rare places where the two peaks occur at the same point (one of these locations is shown in the figure above). This is reminiscent of the wave interference phenomenon in the Young double-slit experiment I described earlier.

If we add yet another wave of slightly different wavelength

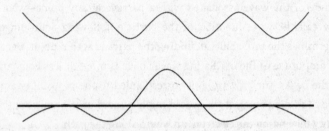

the resulting wave then looks like this:

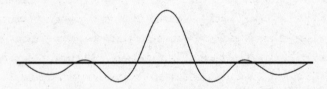

The interference washes out more of the oscillations aside from the position where the two waves line up, making the amplitude of the wave at the peak much higher there than elsewhere.

You can imagine what would happen if I continue this process, continuing to add just the right amount of waves with slightly different frequencies to the original wave. Eventually the resulting wave amplitudes will cancel out more and more at all places except for some small region around the center of the figure, and at faraway places where all the peaks might again line up:

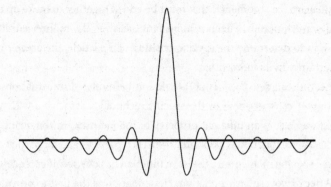

The greater the number of slightly different frequencies that I add together, the narrower will be the width of the largest central peak. Now, imagine that this represents the wave function of some particle. The larger the amplitude of the central peak, the greater the probability of finding the particle somewhere within the width of that peak. But the width of that central peak is still never quite zero, so the disturbance remains spread out over some small, if increasingly narrow, region.

Now recall that Planck and Einstein told us that, for light waves, at least, the energy of each quantum of radiation, i.e., each photon, is directly related to its frequency. Not surprisingly, a similar relation holds for the probability waves associated with massive particles, but in this case it is the momentum of the particle that is related to the frequency of the probability wave associated with the particle.

Hence, Heisenberg's uncertainty relation: If we want to localize a particle over a small region, i.e., have the width of the highest peak in its wave function as narrow as possible, then we must consider that the wave function is made up by adding lots of different waves of slightly different frequencies together. But this means that the momentum of the particle, which is associated with the frequency of its wave function, must be spread out somewhat. The narrower the dominant peak in space in the particle's wave function, the greater the number of different frequencies (i.e., momenta) that must be added together to make up the final wave function. Put in a more familiar way, the more accurately we wish to determine the specific position of a particle, the greater the uncertainty in its momentum.

As you can see, there is no restriction here related to actual observations, *or consciousness*, or the specific technology associated with any observation. It is an inherent property of the fact that, in the quantum world, a wave function is associated with each particle, and for particles of a fixed specific momentum, the wave function has one specific frequency.

After discovering this relation, Heisenberg was the first to provide a heuristic picture of why this might be the case, which he posed in terms of a thought experiment. To measure the position of a particle you have to bounce light off the particle, and to resolve the position with great precision requires light of a wavelength small enough to resolve this position. But the smaller the wavelength, the bigger the frequency and the higher the energy associated with the quanta of that radiation. But bouncing light with a higher and higher energy off the particle clearly changes the particle's energy and momentum. Thus, after the measurement is made, you may know the position of the particle at the time of the measurement, but the range of possible energies and momenta you have imparted to the particle by scattering light off it is now large.

For this reason, many people confuse the Heisenberg uncertainty relation with the "observer effect," as it has become known, in quantum mechanics. But, as the example I have given should demonstrate, inherently the Heisenberg uncertainty principle has nothing to do with observation at all. To paraphrase a friend of mine, if consciousness had anything to do with determining the results of quantum physics experiments, then in reporting the results of physics experiments we would have to discuss what the experimenter was thinking about—for example, sex—when performing the experiment. But we don't. The supernova explosions that produced the atoms that make up your body and mine occurred quite nicely long before our consciousness existed.

The Heisenberg uncertainty principle epitomizes in many ways the complete demise of our classical worldview of nature. Independent of any technology we might someday develop, nature puts an absolute limit on our ability to know, with any degree of certainty, both the momentum and position of any particle.

But the issue is even more extreme than this statement implies. *Knowing* has nothing to do with it. As I described in the earlier double-slit experiment example, there is no sense in which the particle has at any time both a specific position and a specific momentum. It possesses a wide range of both, *at the same time*, until we measure it and thereby fix at least one of them within some small range determined by our measurement apparatus.

· · ·

Following Heisenberg, the next step in unveiling the quantum craziness of reality was taken by an unlikely explorer, Paul Adrien Maurice Dirac. In one sense, Dirac was the perfect man for the job. As Einstein is reputed to have later said of him, "This balancing on the dizzying path between genius and madness is awful."

When I think of Dirac, an old joke comes to mind. A young child has never spoken and his parents go to see numerous doctors to seek help, to no avail. Finally, on his fourth birthday he comes down for breakfast and looks up at his parents and says, "This toast is cold!" His parents nearly burst with happiness, hug each other, and ask the child why he has never before spoken. He answers, "Up to now, everything was fine."

Dirac was notoriously laconic, and a host of stories exist about his unwillingness to engage in any sort of repartee, and also about how he seemed to take everything that was said to him literally. Once, while Dirac was writing on a blackboard during one of his lectures, someone in the audience was reputed to have raised his hand and said, "I don't understand that particular step you have just written down." Dirac stood silent for the longest while until the audience member asked if

Dirac was going to answer the question. To which Dirac said, "There was no question."

I actually spoke to Dirac, one day, on the phone—and I was terrified. I was still an undergraduate and wanted to invite him to a meeting I was organizing for undergraduates around the country. I made the mistake of calling him right after my quantum mechanics class, which made me even more terrified. After a rambling request that I blurted out, he was silent for a moment, then gave a simple one-line response: "No, I don't think I have anything to say to undergraduates."

Personality aside, Dirac was anything but timid in his pursuit of a new Holy Grail: a mathematical formulation that might unify the two new revolutionary developments of the twentieth century, quantum mechanics and relativity. In spite of numerous efforts since Schrödinger (who derived his famous wave equation during a two-week tryst in the mountains with several of his girlfriends), and since Heisenberg had revealed the basic underpinning of quantum mechanics, no one had been successful at fully explaining the behavior of electrons bound deep inside atoms.

These electrons have, on average, velocities that are a fair fraction of the speed of light, and to describe them, we must use Special Relativity. Schrödinger's equation worked well to describe the energy levels of electrons in the outer parts of simple atoms such as hydrogen, where it provided a quantum extension of Newtonian physics. It was not the proper description when relativistic effects needed to be taken into account.

Ultimately Dirac succeeded where all others had failed, and the equation he discovered, one of the most important in modern particle physics, is, not surprisingly, called the Dirac equation. (Some years later, when Dirac first met the physicist Richard Feynman, whom we shall come to shortly, Dirac said after another awkward silence, "I have an equation. Do you?")

Dirac's equation was beautiful, and as the first relativistic treatment of the electron, it allowed correct and precise predictions for the energy levels of all electrons in atoms, the frequencies of light they emit, and thus the nature of all atomic spectra. But the equation had a fundamental problem. It seemed to predict new particles that didn't exist.

To establish the mathematics necessary to describe an electron moving at relativistic speeds, Dirac had to introduce a totally new formalism that used four different quantities to describe electrons.

As far as we physicists can discern, electrons are microscopic point particles of essentially zero radius. Yet in quantum mechanics they nevertheless behave like spinning tops and therefore have what physicists call angular momentum. Angular momentum reflects that once objects start spinning, they will not stop unless you apply some force as a brake. The faster they are spinning, or the more massive they are, the greater the angular momentum.

There is, alas, no classical way of picturing a pointlike object such as an electron spinning around an axis. Spin is thus one of the areas where quantum mechanics simply has no intuitive classical analogue. In Dirac's relativistic extension of Schrödinger's equation, electrons can possess only two possible values for their angular momentum, which we simply call their spin. Think of electrons as either spinning around one direction, which we can call up, or spinning around the opposite direction, which we can call down. Because of this, two quantities are needed to describe the configurations of electrons, one for spin-up electrons and one for spin-down electrons.

After some initial confusion, it became clear that the other two quantities that Dirac needed to describe electrons in his relativistic formulation of quantum mechanics seemed to describe something crazy—*another* version of electrons with the same mass and spin but with the opposite electric charge. If, by convention, electrons have a negative charge, then these new particles would have a positive charge.

Dirac was flummoxed. No such particle had ever been observed. In a moment of desperation, Dirac supposed that perhaps the positively charged particle described by his theory was actually the proton, which, however, has a mass two thousand times larger than that of the electron. He gave some hand-waving arguments for why the positively charged particle might get a heavier mass. The larger weight could be caused by different possible electromagnetic interactions it had with otherwise empty space, which he envisaged might be populated with a possibly infinite sea of unobservable particles. This is actually not as crazy as it sounds, but to describe why would force us toward one of those twists and turns that we want to avoid here. In any case, it was quickly shown that this idea didn't hold water—first, because the mathematics didn't support this argument, and the new particles would have to have the same mass as electrons. Second, if the proton and the electron were in some sense mirror images, then they could annihilate each other so that neutral matter could not be stable. Dirac had to admit that if his theory was true, some new positive version of the electron had to exist in nature.

Fortunately for Dirac, within a year of his resigned capitulation, Carl Anderson found particles in cosmic rays that are identical to electrons but have the opposite charge. The positron was born, and Dirac was heard to say, in response to his unwillingness to accept the implications of his own mathematics, "My equation was smarter than I was!" Much later he reportedly gave another reason for not acknowledging the possibility of a new particle: "Pure cowardice."

Dirac's "prediction," even if reluctant, was a remarkable milestone. It was the first time that, purely on the basis of theoretical notions arising from mathematics, a new particle was predicted. Think about that.

Maxwell had "postdicted" the existence of light as a result of his unification of electricity and magnetism. Le Verrier had predicted the existence of Neptune by using observations of anomalies in the orbit of Uranus. But here was a prediction of a new basic feature of the universe based purely on theoretical arguments about nature at its most fundamental scales, with no direct experimental motivation in advance. It may have seemed like a matter of faith, but it wasn't—after all, the proposer didn't actually believe it—and while like faith it proposed an un-observed reality, unlike faith it proposed a reality that could be tested, and it could have been wrong.

The discovery of relativity by Einstein revolutionized our ideas of space and time, and the discoveries by Schrödinger and Heisenberg of the laws of quantum mechanics revolutionized our picture of atoms. Dirac's first combination of the two provided a new window on the hidden nature of matter at much smaller scales. It heralded the beginning of the modern era in particle physics, setting a trend that has continued for almost a century.

First, if the Dirac equation was applied more generally to other particles, and there was no reason to believe it shouldn't be, then not only would electrons have "antiparticles," as they later became known, so would all the other known particles in nature.

Antimatter has become the stuff of science fiction. Starships such as the USS *Enterprise* in *Star Trek* are invariably powered by antimatter, and the possibility of an antimatter bomb was the silliest part of the plot in the recent mystery thriller *Angels & Demons*. But antimatter *is* real. Not only was the positron discovered in cosmic rays, but antiprotons and antineutrons were discovered later as well.

At a fundamental level, antimatter is not so strange. Positrons are just like electrons, after all, only with the opposite charge. They do not, as many people think, fall "up" in a gravitational field. Matter and antimatter *can* interact and completely annihilate into pure radiation, which seems sinister. But particle-antiparticle annihilation is just one in a host of new possible interactions of elementary particles that can occur once we enter the subatomic realm. Moreover, one would need a large amount of antimatter to actually annihilate enough matter to even light a lightbulb with the energy produced.

Ultimately, that is why antimatter *is* strange. It is strange because the universe we live in is full of matter, and not antimatter. A universe made of antimatter would seem identical to ours. And a universe made of equal amounts of matter and antimatter—which would surely seem the most sensible universe to begin with—would, unless something happened in the meantime, be boring because the matter and antimatter would have long ago annihilated each other and the universe would now contain nothing but radiation.

Why our world is full of matter and not antimatter remains one of the most interesting issues in modern physics. But recognizing that the real reason why antimatter is strange is simply because you never encounter it once caused me to suggest the following analogy. Antimatter is strange in the same sense that Belgians are strange. They are certainly not intrinsically strange, but if you ever ask in a big auditorium full of people, as I have, for the Belgians to raise their hands, almost no one ever does.

Except when I lectured in Belgium, as I did recently, and where I learned my analogy was not appreciated.

Chapter 8

A WRINKLE IN TIME

*For you are a mist that appears for a little time and
then vanishes.*

—JAMES 4:14

Each hidden connection in nature revealed by science
since the time of Galileo has led physics in new and unexpected directions. The unification of electricity and magnetism revealed the hidden
nature of light. Unifying light with Galileo's laws of motion revealed the
hidden connections between space and time embodied in relativity. The
unification of light and matter revealed the strange quantum universe.
And the unification of quantum mechanics and relativity revealed the
existence of antiparticles.

Dirac's discovery of antiparticles came as a result of his "guessing"
the correct equation to describe the relativistic quantum interactions
of electrons with electromagnetic fields. He had little physical intuition
to back it up, which is one reason why Dirac himself and others were
initially so skeptical of his result. Clarifying the physical imperative for
antimatter came through the work of one of the most important physicists of the latter half of the twentieth century, Richard Feynman.

Feynman could not have been more different from Dirac. While

Dirac was taciturn in the extreme, Feynman was gregarious and a charming storyteller. While Dirac rarely, if ever, intentionally joked, Feynman was a prankster who openly enjoyed every aspect of life. While Dirac was too shy to meet women, Feynman, after the death of his first wife, sought out female companions of every sort. Yet, physics breeds strange bedfellows, and Feynman and Dirac will forever be intellectually linked—once again by light. Together they helped complete the description of the long-sought quantum theory of radiation.

Coming a generation after Dirac, Feynman was in awe of him and spoke of him as one of his physics heroes. Therefore, appropriately, a short 1939 paper that Dirac wrote, in which he suggested a new approach to quantum mechanics, would inspire the work that ultimately won Feynman a Nobel Prize.

Heisenberg and Schrödinger had explained how systems behave quantum mechanically starting with some initial state of the system and calculating how it evolves over time. But, once again, light provides the key to another way to think about quantum systems.

We are accustomed to thinking of light as always going in straight lines. But it doesn't. This is manifest when you view a mirage on a long straight highway on a hot day. The road looks wet way up ahead because light from the sky refracts, bending as it crosses the many successive layers of warm air near the surface of the road, until it heads back up to your eye.

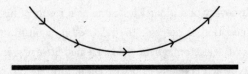

The French mathematician Pierre de Fermat showed in 1650 another way to understand this phenomenon. Light travels faster in warmer, less dense air than it does in colder air. Because the warmest air is near the surface, the light takes less time to get to your eye if it travels down near the ground and then returns up to your eye than it would if it came directly in a straight line to your eye. Fermat formulated a principle, called the Principle of Least Time, which says that, to determine the ultimate trajectory of any light ray, you simply need to examine all possible paths from A to B and find the one that takes the least time.

This makes it sound as if light has intentionality, and I resisted the temptation to say light considers all paths and chooses the one that takes the least time because I fully expect that Deepak Chopra would later quote me as implying that light has consciousness. Light does not have consciousness, but the mathematical result makes it appear as if light chooses the shortest distance.

Now, recall that in quantum mechanics, light rays and electrons do not act as if they take a single trajectory to go from one place to another—they take all possible trajectories at the same time. Each trajectory has a specific probability of being measured, and the classical, least time, trajectory has the largest probability of all.

In 1939, Dirac suggested a way of calculating all such probabilities and summing them to determine the quantum mechanical likelihood that a particle that starts out at A will end up at B. Richard Feynman, as a graduate student, after learning about Dirac's paper at a beer party, mathematically derived a specific example demonstrating that this idea worked. By taking Dirac's hint as a starting point, Feynman derived results that were identical to those that one would derive using the Schrödinger or Heisenberg pictures, at least in simple cases. More important, Feynman could use this new "sum over paths" formula to handle quantum systems that couldn't easily be described or analyzed by the other methods.

Eventually Feynman refined his mathematical technique to help push forward Dirac's relativistic equation for the quantum behavior of electrons and to produce a fully consistent quantum mechanical theory of the interaction between electrons and light. For that work, establishing the theory known as quantum electrodynamics (QED), he shared the Nobel Prize in 1965 with Julian Schwinger and Sin-Itiro Tomonaga.

Even before completing this work, however, Feynman described an intuitive physical reason why relativity, when combined with quantum mechanics, requires the existence of antiparticles.

Consider an electron moving along on a possible "quantum" trajectory. What does this mean? An electron takes all possible trajectories between two points as long as I am not measuring it while it travels. Among these are trajectories that are classically not allowed because they would violate rules such as the limitation that objects cannot travel faster than light (arising from relativity). Now the Heisenberg uncertainty principle says that even if I try to measure the electron along its trajectory over some short time interval, some intrinsic uncertainty in the velocity of the electron remains that can never be overcome. Thus even if I measure the trajectory at various points, I cannot rule out some weird nonclassical behavior during these intervals. Now, imagine the trajectory shown below:

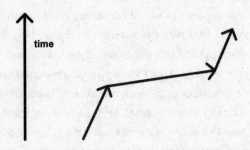

For the short time in the middle of the time interval shown the electron is traveling faster than the speed of light.

But Einstein tells us that time is relative, and different observers will measure different intervals between events. And if a particle is traveling faster than light in one reference frame, in another reference frame it will appear to be traveling backward in time, as shown below (this is one of the reasons relativity restricts all observed particles to travel at speeds less than or equal to the speed of light:

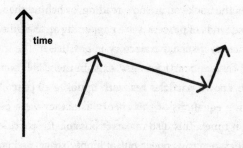

Feynman recognized that in the latter frame this would look like an electron moving forward in time for a little while, then moving backward in time, then moving forward in time. But what does an electron moving backward in time appear like? Since the electron is negatively charged, a negative charge moving backward in time to the right is equivalent to a positive charge moving forward in time to the left. Thus, the picture is equivalent to the following:

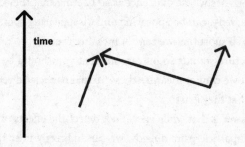

In this picture one starts with an electron moving forward in time, and then sometime later an electron and a particle that appears like an electron but has the opposite charge suddenly appear out of empty space, and the positively charged particle moves to the left, again forward in time, until it encounters the original electron and the two annihilate, leaving only one electron left over to continue moving.

All of this happens on a timescale that cannot be observed directly, for if it could be, then this strange behavior, violating the tenets of relativity, would be impossible. Nevertheless, you can be assured that inside the paper in the book you are now reading, or behind the screen of your ebook, these kinds of processes are happening all the time.

Nevertheless, if such a trajectory is possible in the invisible quantum world, then antiparticles must exist in the visible world—particles identical to known particles but with opposite electric charge (which appear in the equations of this theory as if they were particles going backward in time). This also makes it possible for particle-antiparticle pairs to spontaneously appear out of empty space, as long as they annihilate in a time period quickly enough so that their brief existence cannot be measured.

With this line of reasoning, not only did Feynman give a physical argument for the existence of antiparticles required by the unification of relativity and quantum mechanics, he also demonstrated that at any time we cannot say that only one or two particles are in some region. A potentially infinite number of "virtual" particle-antiparticle pairs— pairs of particles whose existence is so fleeting that they cannot be directly observed—can be appearing and disappearing spontaneously on timescales so short that we cannot measure them.

This picture sounds so outrageous that you should be incredulous. After all, if we cannot measure these virtual particles directly, how can we claim that they exist?

The answer is that while we cannot detect the effects of these virtual particle-antiparticle pairs directly, we can indirectly infer their presence because they can indirectly affect the properties of systems we *can* observe.

The theory in which these virtual particles are incorporated, along with the electromagnetic interactions of electrons and positrons, called quantum electrodynamics, is the best scientific theory we have so far. Predictions based on the theory have been compared with observations, and they agree to more than ten decimal places. In no other area of science can this level of accuracy be obtained in the comparison between observation and prediction, based on the direct applications of fundamental principles on the most basic scales we can describe.

But the agreement between theory and observation is only possible if the effects of virtual particles are included. Indeed, the very phenomenon of virtual particles implies that, in quantum theory, forces between particles are *always* conveyed by the exchange of virtual particles, in a way I shall now describe.

In quantum electrodynamics, electromagnetic interactions occur by the absorption or emission of the quanta of electromagnetism, namely photons. Following Feynman, we can diagram this interaction as an electron emitting a wavy "virtual" photon (γ) and changing direction:

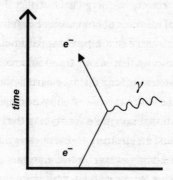

Then, the electric interaction between two electrons can be diagrammed as:

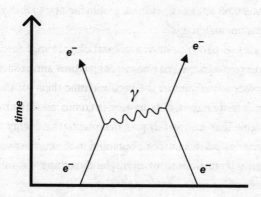

In this case, the electrons interact with each other by exchanging a virtual photon, one that is spontaneously emitted by the electron on the left and absorbed by the other in so short a time that the photon cannot be observed. The two electrons repel each other and move apart after the interaction.

This also explains why electromagnetism is a long-range force. The Heisenberg uncertainty principle tells us that if we measure a system for some time interval, then there is an associated uncertainty in the measured energy of the system. Moreover, as the time interval gets bigger, the associated uncertainty in energy gets smaller. Because the photon is massless, a virtual massless photon, using Einstein's relation between mass and energy, can carry an arbitrarily small amount of energy when it is created. This means that it can travel an arbitrarily long time—and therefore an arbitrarily long distance—before being absorbed, and it will still be protected by the uncertainty principle, as the energy it can carry is so small that no visible violation of the conservation of energy will occur. Thus, an electron on Earth can emit a virtual photon that could travel to Alpha Centauri, four light-years away, and that photon can still produce a force on an electron there that absorbs it. If the photon weren't massless, however, but had some rest mass, m, it would carry with it a minimum energy, given by $E = mc^2$, and could therefore only travel a finite distance (i.e., over a finite time interval) before it would have to be absorbed without producing any visible violation of the conservation of energy.

These virtual particles have a potential problem, however. If one particle can be exchanged or one virtual particle-antiparticle pair can spontaneously appear out of the vacuum, then why not two or three or even an infinite number? Moreover, if virtual particles must disappear in a time that is inversely proportional to the energy they carry, then what stops particles from popping out of empty space carrying an arbitrarily large amount of energy and existing for an arbitrarily small time?

When physicists tried to take into account these effects, they encountered infinite results in their calculations.

The solution? Ignore them.

Actually not ignore them, but systematically sweep the infinite pieces of calculations under the rug, leaving only finite bits left over. This begs the questions of how one knows which finite parts to keep, and why the whole procedure is justified.

The answer took quite a few years to get straight, and Feynman was one of the group who figured it out. But for many years after, including up to the time he won the Nobel Prize in 1965, he viewed the whole effort as a kind of trick and figured that at some point a more fundamental solution would arise.

Nevertheless, a good reason exists for ignoring the infinities introduced by virtual particles with arbitrarily high energies. Because of the Heisenberg uncertainty principle, these energetic particles can propagate only over short distances before disappearing. So how can we be sure that our physical theories, which are designed to explain phenomena at scales we can currently measure, actually operate the same way at these very small scales? Maybe new physics, new forces, and new elementary particles become relevant at very small scales?

If we had to know all the laws of physics down to infinitesimally small scales in order to explain phenomena at the much larger scales we experience, then physics would be hopeless. *We would need a theory of everything before we could ever have a theory of something.*

Instead, reasonable physical theories should be ones that are insensitive to any possible new physics occurring at much smaller scales than the scales that the original theories were developed to describe. We call these theories *renormalizable*, since we "renormalize" the otherwise infinite predictions, getting rid of the infinities and leaving only finite, sensible answers.

Saying that this is required is one thing, but proving that it can be done is something else entirely. This procedure took a long time to get straight. In the first concrete example demonstrating that it made sense, the energy levels of hydrogen atoms were precisely calculated, which allowed a correct prediction of the spectrum of light emitted and absorbed by these atoms as measured in the laboratory.

Although Feynman and his Nobel colleagues elucidated the mechanism to mathematically implement this technique of renormalization, the proof that quantum electrodynamics (QED) was a "renormalizable" theory, allowing precise predictions of all physical quantities one could possibly measure in the theory, was completed by Freeman Dyson. His proof gave QED an unprecedented status in physics. QED provided a complete theory of the quantum interactions of electrons and light, with predictions that could be compared with observations to arbitrarily high orders of precision, limited only by the energy and determination of the theorists doing the calculations. As a result, we can predict the spectra of light emitted by atoms to exquisite precision and design laser systems and atomic clocks that have redefined accuracy in measuring distance and time. The predictions of QED are so precise that we can search in experiments for even minuscule departures from them and probe for possible new physics that might emerge as we explore smaller and smaller scales of distance and time.

With fifty years of hindsight, we now also understand that quantum electrodynamics is such a notable physical theory in part because of a "symmetry" associated with it. Symmetries in physics probe deep characteristics of physical reality. From here on into the foreseeable future, the search for symmetries is what governs the progress of physics.

Symmetries reflect that a change in the fundamental mathematical quantities describing the physical world produce no change in the way the world works or looks. For example, a sphere can be rotated in any direction by any angle, and it still looks precisely the same. Nothing about the physics of the sphere depends on its orientation. That the laws of physics do not change from place to place, or time to time, is of deep significance. The symmetry of physical law with time—that nothing about the laws of physics appears to change with time—results in the conservation of energy in the physical universe.

In quantum electrodynamics, one fundamental symmetry is in the nature of electric charges. What we call "positive" and "negative" are clearly arbitrary. We could change every positive charge in the universe to negative, and vice versa, and the universe would look and behave precisely the same.

Imagine, for example, that the world is one giant chessboard, with black and white squares. Nothing about the game of chess would be changed if I changed black into white, and white into black. The white pieces would become black pieces and vice versa, and otherwise the board would look identical.

Now, precisely because of this symmetry of nature, the electric charge is conserved: no positive or negative charge can spontaneously appear in any process, even due to quantum mechanics, without an equal and opposite charge appearing at the same time. For this reason, virtual particles are only produced spontaneously in empty space in combination with antiparticles. It is also why lightning storms occur on Earth. Electric charges build up on Earth's surface because storm clouds build up large negative charges at their base. The only way to get rid of this charge is to have large currents flow from the ground upward into the sky.

The conservation of charge resulting from this symmetry can be understood using my chessboard analogy. That every white square must be located next to a black square means that whenever I switch black and white, the board ultimately looks the same. If I had two black squares in a row, which would mean the board had some net "blackness," then "black" and "white" would no longer be equivalent arbitrary labels. Black would be physically different from white. In short, the symmetry between black and white on the board would be violated.

Bear with me now, because I am about to introduce a concept that is much more subtle, but much more important. It's so important that essentially all of modern physical theory is based on it. But it's so subtle that without using mathematics, it is hard to describe. It is so subtle that its ramifications are still being unraveled today, more than a hundred years since it was first suggested. So, don't be surprised if it takes one or two readings to fully get your head around the idea. It has taken physicists much of the past century to get their heads around it.

This symmetry is called gauge symmetry for an obscure historical reason I shall describe a bit later. But the strange name is irrelevant. It is what the symmetry implies that is important:

> *Gauge symmetry in electromagnetism says that I can actually change my definition of what a positive charge is locally at each point of space without changing the fundamental laws associated with electric charge, as long as I also somehow introduce some quantity that helps keep track of this change of definition from point to point. This quantity turns out to be the electromagnetic field.*

Let's try to parse this using my chessboard analogy. The global symmetry I described before changes black to white everywhere, so when the chessboard is turned by 180 degrees, it looks the same as it did before and the game of chess is clearly not affected.

Now, imagine instead that I change black to white in one square, and I don't change white to black in the neighboring square. Then the board will have two adjacent white squares. This board, with two adjacent white squares, clearly won't look the same as it did before. The game cannot be played as it was before.

But hold on for a moment. What if I have a guidebook that tells me what game pieces should do every time they encounter adjacent squares where one color has been changed but not the next. Then the rules of the game can remain the same, as long as I consult the guidebook each time I move. *This guidebook therefore allows the game to proceed as if nothing were changed.*

In mathematics, a quantity that ascribes some rule associated with each point on a surface like a chessboard is called a function. In physics, a function defined at every point in our physical space is called a field, such as, for example, the electromagnetic field, which describes how strong electric and magnetic forces are at each point in space.

Now here's the kicker. The properties that must characterize the form of the necessary function (which allows us to change our definition of electric charge from place to place without changing the underlying physics governing the interaction of electric charges) are *precisely* those that characterize the form of the rules governing electromagnetic fields.

Put another way, the requirement that the laws of nature remain invariant under a *gauge transformation*—namely some transformation that locally changes what I call positive or negative charge—identically requires the existence of an electromagnetic field that is governed precisely by Maxwell's equations. *Gauge invariance*, as it is called, completely determines the nature of electromagnetism.

This presents us with an interesting philosophical question. Which is more fundamental, the symmetry or the physical equations that manifest the symmetry? In the former case, where this gauge symmetry of nature requires the existence of photons, light, and all the equations and phenomena first discovered by Maxwell and Faraday, then God's apparent command "Let there be light" becomes identical with the command "Let electromagnetism have a gauge symmetry." It is less catchy, perhaps, but nevertheless true.

Alternatively, one could say that the theory is what it is, and the discovery of a mathematical symmetry in the underlying equations is a happy accident.

The difference between these two viewpoints seems primarily semantic, which is why it might interest philosophers. But nature does provide some guidance. If quantum electrodynamics were the only theory in nature that respected such a symmetry, the latter view might seem more reasonable.

But *every* known theory describing nature at a fundamental scale reflects some type of gauge symmetry. As a result, physicists now tend to think of symmetries of nature as fundamental, and the theories that then describe nature as being restricted in form to respect these symmetries, which in turn then reflect some key underlying mathematical features of the physical universe.

Whatever one might think of regarding this epistemological issue, what matters in the end to physicists is that the discovery and application of this mathematical symmetry, gauge symmetry, has allowed us to discover more about the nature of reality at its smallest scales than any other idea in science. As a result, all attempts to go beyond our current understanding of the four forces of nature, electromagnetism, the two forces associated with atomic nuclei, the strong and weak forces, which we shall meet shortly, and gravity—including the attempt to create a quantum theory of gravity—are built on the mathematical underpinnings of gauge symmetry.

. . .

That gauge symmetry has such a strange name has little to do with quantum electrodynamics and is an anachronism, related to a property of Einstein's General Theory of Relativity, which, like all other fundamental theories, also possesses gauge symmetry. Einstein showed that we are free to choose any local coordinate system we want to describe the space around us, but the function, or field, that tells us how to connect these coordinate systems from point to point is related to the underlying curvature of space, determined by the energy and momentum of material in space. The coupling of this field, which we recognize as the gravitational field, to matter, is precisely determined by the invariance of the geometry of space under the choice of different coordinate systems.

The mathematician Hermann Weyl was inspired by this symmetry of General Relativity to suggest that the form of electromagnetism might also reflect an underlying symmetry associated with physical changes in length scales. He called these different "gauges," inspired by the various track gauges of railroads. (Einstein, and Sheldon on *The Big Bang Theory*, aren't the only physicists who have been inspired by trains.) While Weyl's guess turned out to be incorrect, the symmetry that *does* apply to electromagnetism became known as gauge symmetry.

Whatever the etymology of the name, gauge symmetry has become the most important symmetry we know of in nature. From a quantum perspective—in the quantum theory of electromagnetism, quantum electrodynamics—the existence of gauge symmetry becomes even more important. It is the essential feature that ensures that QED is sensible.

If you think about the nature of symmetry, then it begins to make sense that such a symmetry might ensure that quantum electrodynamics makes sense. Symmetries tell us, for example, that different parts of the natural world are related, and that certain quantities remain the same under various types of transformations. A square looks the same when we rotate it ninety degrees because the sides are all the same length and the angles at each corner are the same. So, symmetry can tell us that different mathematical quantities that result from physical calculations, such as the effects of many virtual particles, and many virtual antiparticles, for example, can have the same magnitude. They may also have opposite signs so that they might cancel *exactly*. The existence of this symmetry is what can require such exact cancellations.

In this way, one might imagine that in quantum electrodynamics the nasty terms that might otherwise give infinite results can cancel with other potentially nasty terms, and all the nastiness can disappear. And this is precisely what happens in QED. The gauge symmetry ensures that any infinities that might otherwise arise in deriving physical predictions can be isolated in a few nasty terms that can be shown by the symmetry to either disappear or to be decoupled from all physically measurable quantities.

This profoundly important result, proven by decades of work by some of the most creative and talented theoretical physicists in the world, established QED as the most precise and preeminent quantum theory of the twentieth century.

Which made it all the more upsetting to discover that, while this mathematical beauty indeed allowed a sensible understanding of one of nature's fundamental forces—electromagnetism—other nastiness began when considering the forces that govern the behavior of atomic nuclei.

Chapter 9

DECAY AND RUBBLE

There is no new thing under the sun.

—ECCLESIASTES 1:9

When I first learned that we human beings are radioactive, it shocked me. I was in high school listening to a lecture by the remarkable polymath and astrophysicist Tommy Gold, who had done pioneering work in cosmology, pulsars, and lunar science, and he informed us that the particles that made up most of the mass of our bodies, neutrons, are unstable, with a mean lifetime of about ten minutes.

Given, I hope, that you have been reading this book for longer than ten minutes, this may surprise you too. The resolution of this seeming paradox is one of the first and most wonderful of the gorgeous accidents of nature that make our existence possible. As we continue to explore more deeply the question "Why are we here?," this accident will loom large on the horizon. While the neutron may seem far removed from light, which has been the centerpiece of our story thus far, we shall see that the two are ultimately deeply connected. The decay of neutrons—responsible for the "beta decay" of unstable nuclei—required physicists to move beyond their simple and elegant theories of light and open up new fundamental areas of the universe for investigation.

But I am getting ahead of myself.

In 1929, when Dirac first wrote down his theory of electrons and radiation, it looked as if it might end up being a theory of almost everything. Aside from electromagnetism, the only other force in town was gravity, and Einstein had just made great strides in understanding it. Elementary particles consisted of electrons, photons, and protons, together comprising all the objects that appeared necessary to understand atoms, chemistry, life, and the universe.

The discovery of antiparticles upset the applecart somewhat, but since Dirac's theory had effectively predicted them (even if Dirac himself had to catch up with the theory), this was more like a speed bump on the road to reality than a roadblock or detour.

Then came 1932. Up to that time, scientists had presumed that atoms were composed entirely of protons and electrons. This posed a bit of a problem, however, because the masses of atoms didn't quite add up. In 1911 Rutherford discovered the existence of the atomic nucleus, containing almost all the mass of atoms in a small region one hundred thousand times smaller than the size of the orbits of the electrons. Following that discovery, it became clear that the mass of heavy nuclei was just a bit more than twice the mass that could be accounted for if the number of protons in the nucleus equaled the number of electrons orbiting the nucleus, ensuring that atoms would be electrically neutral.

The proposed solution to this conundrum was simple. Actually twice as many protons were in the nucleus as electrons surrounding it, but just the right number of electrons were trapped inside the nucleus, so that again the total electric charge of the atom would be equal to zero.

However, quantum mechanics implied that the electrons couldn't be confined within the nucleus. The argument is a bit technical, but it goes something like this: If elementary particles have a wavelike character, then if one is going to confine them to a small distance, the magnitude of their wavelength must be smaller than the confinement scale. But the wavelength associated with a particle is, in quantum mechanics, inversely proportional to the momentum carried by the particle, and hence also inversely proportional to the energy carried by the particle. If electrons were confined to a region the size of an atomic nucleus, the energy they would need to possess would be about a million times the energy associated with the characteristic energies released by electrons as they jump between energy levels in their atomic orbits.

How could they achieve such energies? They couldn't. For, even if electrons were tightly bound to protons within nuclei by electronic forces, the binding energy that would be released in this process as they "fell" into the nucleus would be more than ten times smaller than the energy needed to confine the quantum mechanical electron wave function to a region contained within the nucleus.

Here too the numbers just didn't add up.

Physicists at the time were aware of the problem, but lived with it. I suspect that an agnostic approach was deemed prudent, and physicists were willing to suspend disbelief until they knew more, because the issues involved the cutting-edge physics of quantum mechanics and atomic nuclei. Instead of proposing exotic new theories (there were probably some at the margins that I am not aware of), the community was eventually driven by experiments to overcome its natural hesitation to take the logical next step: to assume nature was more complicated than had thus far been revealed.

In 1930, about the time that Dirac was coming to grips with the possibility that his antiparticles weren't really protons, a series of experiments provided just the clues that were needed to unravel the nuclear paradox. The poetry of the discoveries was rivaled only by the drama in the private lives of the researchers.

Max Planck had helped pioneer the quantum revolution by resolving the paradox of the spectrum of radiation emitted by atomic systems. So it was fitting that Planck should indirectly help resolve the paradoxical makeup of the nucleus. While he didn't himself spearhead the relevant research, he recognized the talents of a young student of mathematics, physics, chemistry, and music at the University of Berlin, Walther Bothe, and in 1912 Planck accepted him as a doctoral student and mentored him throughout the rest of his career.

Bothe was spectacularly lucky to be mentored by Planck and, shortly thereafter, by Hans Geiger, of Geiger counter fame. Geiger, in my mind, is one of the most talented experimental physicists to have been overlooked for a Nobel Prize. Geiger had begun his career by doing the experiments, with Ernest Marsden, that Ernest Rutherford utilized to discover the existence of the atomic nucleus. Geiger had just returned from England, where he'd worked with Rutherford, to direct a new laboratory in Berlin, and one of his first acts was to hire Bothe as an assistant. There Bothe learned to focus on important experiments, using simple approaches that yielded immediate results.

After an "involuntary vacation" of five years, as a prisoner of war in Siberia during the First World War, Bothe returned and built a remarkable collaboration with Geiger, eventually succeeding him as director of the laboratory. During their time together they pioneered the use of "coincidence methods" to explore atomic, and eventually nuclear, physics. Using different detectors located around a target, and using careful timing, they could look for simultaneous events, signaling that the source had to be a single atomic or nuclear decay.

In 1930 Bothe and his assistant Herbert Becker observed something completely new and unexpected. While bombarding beryllium nuclei with products of nuclear decay called alpha particles (already known to be the nuclei of helium), the two observed the emission of a completely new form of high-energy radiation. This radiation had two unique features. It was more penetrating than the most energetic gamma rays, but like gamma rays, the radiation was composed of electrically neutral particles so that it did not ionize atoms as it passed through matter.

News of this surprising discovery made its way to other physics laboratories throughout Europe. Bothe and Becker had initially proposed that this radiation was some *new* sort of gamma ray. In Paris, Irène

Joliot-Curie, the daughter of famed physicist Marie Curie, and Irène's husband, Frédéric, replicated Bothe and Becker's results and explored the radiation in more detail. In particular, they found that when it bombarded a paraffin target, it knocked out protons with incredible energy.

This observation made it clear that the radiation couldn't be a gamma ray. Why?

The answer is relatively simple. If you throw a piece of popcorn at an oncoming truck, you are unlikely to stop the truck or even break a window. That is because the popcorn, even if you throw it with great energy, carries little momentum because the popcorn is light. To stop a truck you have to change its momentum by a large amount because, even if it is moving slowly, it is heavy. To stop a truck or knock a heavy object off the truck, you have to throw a big rock.

Similarly, to knock out a heavy particle such as a proton from paraffin, a gamma ray, made of massless photons, would have to carry great energy (so that the momentum carried by the individual photons was large enough to kick out a heavy proton), and not enough energy was available, by an order of magnitude at least, in any known nuclear-decay processes for this.

Surprisingly, the Joliot-Curies (they were modern and both adopted the same hyphenated last name) were probably loath, like Dirac, to propose new elementary particles to explain data—since protons, electrons, and photons were not only familiar, but sufficient up to that time to explain everything known, including exotic quantum phenomena associated with atoms. So, Irène and Frédéric didn't make the now-obvious proposal that maybe a new neutral massive particle was being produced in the decays that Bothe and Becker had discovered. Unfortunately, a similar timidity caused the Joliot-Curies to fail to claim discovery of the positron—in spite of having actually observed it in their experiments before Carl Anderson reported his own discovery somewhat later.

It fell to the physicist James Chadwick to push things further. Chadwick clearly had a great nose for physics, but his political acumen was not so sharp. After graduation from the University of Manchester with a master's degree in 1913, working with Rutherford, he obtained a fellowship that would allow him to study anywhere. So he went to Berlin to work with Geiger. He couldn't have picked a better mentor, and he began to do important studies of radioactive decays. Unfortunately, the First World War broke out while Chadwick was in Germany, and he spent the next four years in an internment camp.

Eventually he returned to Cambridge, where Rutherford had since moved, to complete his PhD under Rutherford's direction. Following this Chadwick stayed on to work with Rutherford and help direct the Cavendish laboratory there. While he was aware of Bothe and Becker's results and even reproduced them, only when one of his students informed him of the Joliot-Curies' results did Chadwick become convinced, using the energy argument I mentioned above, that the radiation that had been observed had to result from a new neutral particle—of mass comparable to that of the proton—that might reside in atomic nuclei, an idea he and Rutherford had been germinating for years.

Chadwick reproduced and extended the Joliot-Curies' experiments, bombarding targets other than paraffin to explore the outgoing protons. He confirmed not only that the energetics of the collisions made it impossible for the source to be gamma rays, but also that the interaction strength of the new particles with nuclei was far greater than would be predicted for gamma rays.

Chadwick didn't dawdle. Within two weeks of beginning his experiments in 1932, he sent a letter to *Nature* entitled "Possible Existence of a Neutron" and followed this up with a more detailed article sent to the Royal Society. The neutron, which we now know makes up most of the mass of heavier nuclei, and thus most of the mass in our bodies, had been discovered.

For his discovery he was awarded the Nobel Prize in Physics three years later, in 1935. In a kind of poetic justice, three of the people whose experiments had made Chadwick's results possible—but who missed out on identifying the neutron—were awarded Nobel Prizes for other work. Bothe won the Nobel Prize in 1954 for his work on using coincidences between observed events in different detectors to explore the detailed nature of nuclear and atomic phenomena. Both Irène and Frédéric Joliot-Curie, who barely missed out on two other Nobel Prize–winning discoveries, won the Nobel Prize in Chemistry in 1935 for their discovery of artificial radioactivity—which was later an essential ingredient in the development of both nuclear power and nuclear weapons. Interestingly, only after winning the Nobel Prize was Irène awarded a professorship in France. With the two Nobel Prizes for her mother, Marie, the Curie family garnered a total of five Nobel Prizes, the most that have ever been received by a single family.

After his discovery Chadwick set out to measure the mass of the neutron. His first estimate, in 1933, suggested a mass of slightly less than the sum of the masses of a proton and an electron. This reinforced the idea that perhaps the neutron was a bound state of these two particles, and the mass difference, using Einstein's relation $E = mc^2$, was due to the energy lost in binding them together. However, after several conflicting measurements by other groups, further analysis a year later by Chadwick using a nuclear reaction induced by gamma rays—which allowed all energies to be measured with great precision—definitely indicated that the neutron was heavier than the sum of the proton and electron masses, even if barely so, with the mass difference being less than 0.1 percent.

It is said that "close" only matters when tossing horseshoes or hand grenades, but the closeness in mass between the proton and the neutron matters a great deal. It is one of the key reasons we exist today.

Henri Becquerel discovered radioactivity in uranium in 1896, and only three years later Ernest Rutherford discerned that radioactivity occurred in two different types, which he labeled alpha and beta rays. A year later gamma rays were discovered, and Rutherford confirmed them as a new form of radiation in 1903, when he gave them their name. Becquerel determined in 1900 that the "rays" in beta decay were actually electrons, which we now know arise from the decay of the neutron.

In beta decay a neutron splits into a proton and an electron, which, as I describe below, would not be possible if the neutron weren't slightly heavier than protons. What is surprising about this neutron decay is not that it occurs, but that it takes so long. Normally the decay of unstable elementary particles occurs in millionths or billionths of a second. Isolated neutrons live, on average, more than ten minutes.

One of the chief reasons that neutrons live so long is that the mass of the neutron is only slightly more than the sum of the masses of a proton plus an electron. Thus, there is only barely enough energy available, via the neutron's rest mass, to allow it to decay into these particles and still conserve energy. (The other reason is that a neutron doesn't decay into only a proton plus an electron. It decays into three particles ... stay tuned!)

While ten minutes may be an eternity on atomic timescales, it is pretty short compared to a human life or the lifetime of atoms on Earth. Returning to the puzzle I mentioned at the beginning of this chapter, what gives? How can we be largely made up of neutrons if they decay before the first commercial break in a thirty-minute TV show?

The answer again lies in the extreme closeness of the neutron and proton masses. A free neutron decays in ten minutes or so. But consider a neutron bound inside an atomic nucleus. Being bound means that it takes energy to kick it out of the nucleus. But that means that it loses energy when it gets bound to the nucleus in the first place. But, Einstein told us that the total energy of a massive particle is proportional to its mass, via $E = mc^2$. That means that, if the neutron loses energy when it gets bound in a nucleus, its mass gets smaller. But since its mass when it is isolated is just a smidgen more than the sum of the masses of a proton and an electron, when it loses mass, it no longer has sufficient energy to decay into a proton and an electron. If it were to decay into a proton, it would have to either release enough energy to also eject the proton from the nucleus, which, given standard nuclear-binding energies, it would not have, or else release enough energy to allow the new proton to re-main in a new stable nucleus. Since the new nucleus would be that of a different element, adding one additional positive charge to the nucleus also generally requires more energy than the minute amount available when a neutron decays. As a result, the neutron and most atomic nuclei containing neutrons remain stable.

The entire stability of the nuclei that make up everything we see, including most of the atoms in our body, is an accidental consequence of the fact that the neutron and proton differ in mass by only 0.1 percent, so that a small shift in the mass of the former, when embedded in nuclei, means it can no longer decay into the latter. That is what I learned from Tommy Gold.

It still amazes me when I think about it. The existence of complex matter, the periodic table, everything we see, from distant stars to the keyboard I am typing this on—hinges on such a remarkable coincidence. Why? Is it an accident, or do the laws of physics require it for some unknown reason? Questions such as these drive us physicists to search deeper for possible answers.

The discovery of the neutron, and the subsequent observation of its decay, introduced more than one new particle into the subatomic zoo. It suggested that perhaps two of the most fundamental properties of nature—the conservation of energy and the conservation of momentum—might break down on the microscopic-distance scales of nuclei.

Almost twenty years before discovering the neutron, James Chadwick had observed something strange about beta rays, well before he or anyone else knew that they originated from decaying neutrons. The spectrum of energy carried by electrons emitted in neutron decay is continuous, going from essentially zero energy up to a maximum energy, which depends on the energy available after the neutron has decayed—for a free neutron this maximum energy is the energy difference between the mass of the neutron and the sum of the masses of the proton and electron.

There is a problem with this, however. It is easiest to see the problem if we imagine for the moment that the proton and the electron have equal masses. Then, if the proton carries off more energy than the electron after the decay, it would be moving faster than the electron. But if they have the same mass, then the proton would also have more momentum than the electron. But if the neutron decays at rest, then its momentum before the decay would be zero, so the momentum of the outgoing proton would have to cancel that of the outgoing electron. But that is impossible unless they have equal momenta, going in opposite directions. So the magnitude of the proton's momentum could never be greater than that of the electron. In short, there is only one value for the energy and the momentum of the two particles after the decay if they have equal masses.

The same reasoning, though mathematically a bit more involved, applies even if the proton and electron have different masses. If they are the only two particles produced in the decay of the neutron, their speeds, and hence their energy and momenta, would be required to each have unique, fixed values that depend on the ratio of their respective masses.

As a result, if electrons from beta decay of neutrons come off with a range of different energies, this would violate the conservation of energy and momentum. But, as I subtly suggested above, this is only true if the electron and proton are the only particles produced as products of the neutron decay.

Again, in 1930, only a few years before the discovery of the neutron, the remarkable Austrian theoretical physicist Wolfgang Pauli wrote a letter to colleagues at the Swiss Federal Institute of Technology, beginning with the immortal header "Dear radioactive ladies and gentlemen," in which he outlined a proposal to resolve this problem, which he also said he didn't "feel secure enough to publish anything about." He proposed that a new electrically neutral elementary particle existed, which he called a neutron, and that in addition to the electron and the proton this new neutral particle was produced in beta decay so that the electron, proton, and this particle together could share the energy available in the decay, allowing a continuous spectrum.

Pauli, who later won the Nobel Prize for his "exclusion principle" in quantum mechanics, was no fool. In fact, he had no patience for fools. He was famous for supposedly rushing up to the blackboard during lectures and removing the chalk from the speaker's hand if he felt nonsense was being spouted. He could be scathingly critical of theories he didn't like, and his worst criticism was reserved for any idea that was so vague, as he put it, "it isn't even wrong." (A dear old colleague of mine when I taught at Yale, the distinguished mathematical physicist Feza Gürsey, once responded to a reporter who asked what was the significance of an announcement of some overhyped idea proposed by some scientists seeking publicity by saying, "It means Pauli must be dead.")

Pauli realized that proposing a new elementary particle that hadn't been observed was speculative in the extreme, and he argued in his letter that such a particle was unlikely both because it had never been seen and would therefore have to interact weakly with matter, and also because it would have to be very light to be produced along with an electron, given that the energies available in beta decay were so small compared to the proton's mass.

The first problem that arose with his idea was the name he chose. After Chadwick's 1932 experimental discovery of the particle we now call the neutron, appropriate for a neutral cousin of the proton with comparable mass, Pauli's hypothesized particle needed another name. The brilliant Italian physicist and colleague of Pauli's—Enrico Fermi— came up with a solution in 1934, changing its name to *neutrino*, an Italian pun for "little neutron."

It would take twenty-six years for Pauli's neutrino to be discovered, enough time for the little particle, and its heavier cousin, the neutron, to force physicists to totally revamp their views on the forces that govern the cosmos, the nature of light, and even the nature of empty space.

Chapter 10

FROM HERE TO INFINITY: SHEDDING LIGHT ON THE SUN

I have fought a good fight, I have finished my
course, I have kept the faith.

—2 TIMOTHY 4:7

The physicist Enrico Fermi is largely unsung in the public's eyes, but he remains one of the greatest twentieth-century physicists. He, together with Richard Feynman, more than any of the other remarkable figures from that equally remarkable period in physics, most influenced my own attitude and approach to the field, as well as my own understanding of it. I only wish I were as talented as either of them.

Born in 1901, Fermi died at the age of fifty-three of cancer, perhaps brought on by his work on radioactivity. In 1954, when he died, he was nine years younger than I am as I write this. But in his short life he pushed forward the frontiers of both experimental and theoretical physics in a way that no one has since repeated, and no one is ever likely to do again. The complexity of the array of theoretical tools now used to develop physical models, and the complexity of machinery now used to test them, are separately too sophisticated to allow any single individual today, no matter how talented, to remain on the vanguard of both endeavors at the level Fermi achieved in his time.

In 1918, when Fermi graduated from high school in Rome, the possibilities open to a brilliant young scientific mind were far less constrained. Quantum mechanics had just been born, new ideas were everywhere, and the rigorous mathematics necessary to deal with these ideas had not yet been developed or applied. Experimental physics had yet to enter the domain of "big science"; experiments could be performed by individual researchers in makeshift laboratories, and they could be completed in weeks instead of months.

Fermi applied to the prestigious Scuola Normale Superiore in Pisa, which required an essay as part of the entrance exam. The theme that year was "specific characteristics of sounds." Fermi submitted an "essay" that included solving partial differential equations for a vibrating rod and applying a technique called Fourier analysis. Even today, these mathematical techniques are not normally encountered until maybe the third year of an undergraduate degree, and for some students not until graduate school. But as a seventeen-year-old, Fermi sufficiently impressed the examiners to receive first place in the exam.

At the university, Fermi first majored in mathematics but switched to physics and largely taught himself General Relativity—which Einstein had only developed a few years earlier—as well as quantum mechanics and atomic physics, which were then emerging fields of research. Within three years of arriving at the university he published theoretical papers in major physics journals on subjects from General Relativity to electromagnetism. At the age of twenty-one, four years after beginning his university studies, he received his doctoral degree for a thesis exploring the applications of probability to X-ray diffraction. At the time a thesis on purely theoretical issues was not acceptable for a physics doctorate in Italy, so this encouraged Fermi to ensure his competence in the laboratory as well as with pen and paper.

Fermi moved to Germany, the center of the emerging research on quantum mechanics, and then to Leiden, Holland, where he met with the most famous physicists of the day—Born, Heisenberg, Pauli, Lorentz, and Einstein, to name a few—before returning to Italy to teach. In 1925, Wolfgang Pauli proposed the "exclusion principle," which disclosed that two electrons could not occupy exactly the same quantum state at the same time and place, and which laid the basis of all of atomic physics. Within a year, Fermi applied this idea to systems of many such identical particles that, like electrons, have two possible values of spin, angular momentum, which we call spin up, and spin down. He thus established the modern form of the field called statistical mechanics, which is at the basis of almost all materials science, semiconductors, and those areas of physics that led to the creation of modern electronic components such as computers.

As I earlier emphasized, there is no intuitive way to picture a point particle as spinning around some axis. It is simply one of the ways that quantum mechanics evades our notions of common sense. Electrons are called spin ½ particles because the magnitude of their spin angular momentum turns out to be half as big as the lowest value of angular momentum associated with the orbital motion of electrons in atoms. Any spin ½ particle such as an electron is called a fermion, named in Fermi's honor.

At the tender age of twenty-six Fermi was elected to a new chair in theoretical physics at the University of Rome and thereafter led a vibrant group of students, including several subsequent Nobel laureates, as they explored atomic and then nuclear physics.

In 1933, Fermi was motivated by another proposal of Pauli's, that for the new particle produced in the decay of neutrons, which Fermi labeled a neutrino. But naming the new particle was just an aside. Fermi had much bigger fish to fry, and he produced a theory for neutron decay that revealed the possible existence of a new fundamental force in nature, the first new force known to science beyond electromagnetism and gravity—which was in its own way inspired by thinking about light. Although it wasn't obvious at the time, this was to be the first of two new forces associated with atomic nuclei, which together with electromagnetism and gravity, comprise all the forces known to operate in nature, from the smallest subatomic scales to the motion of galaxies.

When Fermi submitted his proposal to the journal *Nature*, the editor turned it down because it was "too remote from physical reality to be of interest to readers." For many of us who have since had papers rejected by equally high-handed editors at that journal, it is comforting to know that Fermi's paper, one of the most important proposals in twentieth-century physics, also didn't make the cut.

This inappropriate rejection was undoubtedly frustrating to Fermi, but it did have a useful side effect. Fermi decided instead to return to experimental physics, and in short order he began to experiment with the neutrons discovered by Chadwick two years earlier. Within several months Fermi had developed a powerful radioactive source of neutrons and found that he was able to induce radioactive decays in otherwise stable atoms by bombarding them with neutrons. Bombarding uranium and thorium with neutrons, he also witnessed nuclear decays and thought he had created new elements. In fact, he had actually caused the nuclei to split, or fission, into lighter nuclei, which were later found to also emit more neutrons than they absorbed in the process—as other scientists discovered in 1939.

Fermi's segue into experiment turned out to be good for him. Four years later, in 1938, at the age of thirty-seven, he was awarded the Nobel Prize for introducing artificial radioactivity, creating new radioactive elements by neutron bombardment. Yet by 1938 the Nazis had begun to establish their racial laws in Germany, and Italy had followed suit, so Fermi's Jewish wife, Laura, was endangered. So, after receiving the prize in Stockholm, Fermi and his family didn't return to Italy but moved to New York City, where he accepted a position at Columbia.

When Fermi learned the news about nuclear fission in 1939 in New York, following a lecture by Niels Bohr at Princeton, Fermi amended his earlier Nobel acceptance speech to clarify his earlier error and in short order reproduced the German results. Before long, he and his collaborators realized that this produced the possibility of a chain reaction. Neutrons could bombard uranium, causing it to fission and release energy, and to release more neutrons that could bombard more uranium atoms and so on.

Soon after, Fermi gave a lecture to the US Navy warning of the potential significance of this result, but few took him seriously. Later that year, Einstein's famous letter made its way to President Roosevelt and changed the course of history.

Fermi had recognized the potential dangers inherent in releasing the energy of the atomic nucleus even earlier. A year after getting his doctorate, in 1923, he wrote the appendix for a book on relativity and talked of the potential of $E = mc^2$, writing at the time, "It does not seem possible, at least in the near future, to find a way to release these dreadful amounts of energy—which is all to the good because the first effect of an explosion of such a dreadful amount of energy would be to smash into smithereens the physicist who had the misfortune to find a way to do it."

That idea must have been on his mind in 1941 when, as part of the newly established Manhattan Project, Fermi was assigned the task of creating a controlled chain reaction—namely creating a nuclear reactor. While those in charge were understandably worried about doing this in an urban area, Fermi was confident enough to convince the leader of the project to allow him to build it at the University of Chicago. On December 2, 1942, the reactor went critical, and Chicago survived.

Two and a half years later, Fermi was on hand in New Mexico to observe the first nuclear explosion, the Trinity test. Typical of Fermi, while the others stood in awe and horror, he conducted an impromptu experiment to estimate the bomb's strength by dropping several strips of paper when the blast wave came by, to see how far they were carried.

Fermi's constant experimental approach to physics is one of the reasons I cherish his memory. He always found a simple, easy way to reach the correct answer. Even though he had great mathematical skill, he disliked complication, and he realized that he could get an approximate answer that was "good enough" in a short time, while getting the exact answer might take months or years. He refined his abilities and helped his students do so by inventing what we now call Fermi Problems, which he is also said to have assigned at lunchtime each day to the team working for him. My favorite problem, which I always assign to my introductory-physics students, is "How many piano tuners are there in Chicago?" Try it. If you get between one hundred and five hundred, you did well.

Fermi won the Nobel Prize for his experimental work, but his theoretical legacy for physics may be far greater. True to form, the "theory" he proposed in his famously rejected paper on neutron decay was remarkably simple, yet it did the job. It wasn't a full theory at all, and at the time it would have been premature to develop one. Instead he made the simplest possible assumption. He imagined some new kind of interaction between particles that took place at a single point. The four particles were a neutron, a proton, an electron, and the new particle Pauli and Fermi named the neutrino.

The starting point of Fermi's thinking involved light, as did almost all of modern physics, and in this case the modern quantum theory of light interacting with matter. Recall that Feynman developed a pictorial framework to think about fundamental processes in space and time, when he argued that antimatter should exist. The space-time picture of an electron emitting a photon is reproduced here, but with the electron replaced by a proton, p:

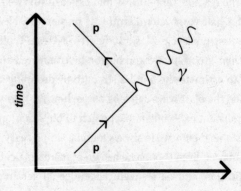

Fermi imagined the decay of a neutron in a similar fashion, but instead of the neutron emitting a photon and remaining the same particle, the neutron, n, would emit a pair of particles—an electron, e, and a neutrino, ν, and would be converted into a proton, p:

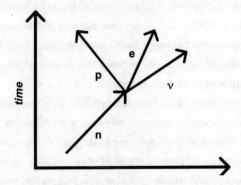

In electromagnetism the strength of the interaction between charged particles and photons (determining the probability of emitting a photon at the point shown in the first figure on the previous page) is proportional to the charge of the particle. Since the charge is what allows particles to interact, or "couple" to the electromagnetic field, we call the magnitude of the fundamental quantum of charge—the charge on a single electron or proton—the "coupling constant" of electromagnetism.

In Fermi's interaction the numerical quantity that appears at the interaction point in the figure where a neutron converts into a proton determines the probability of such a conversion. The value of this quantity is determined by experiment, and we now call it the Fermi constant. Relative to electromagnetism, the numerical value of this quantity is small because the neutron takes a long time to decay—compared, for example, to the rate at which electromagnetic transitions take place in atoms. As a result, Fermi's interaction, describing a new force in nature, became known as the weak interaction.

One of the things that made Fermi's proposal so remarkable was that it was the first time in physics that anyone had proposed that particles other than photons could be spontaneously created in the quantum world. (In this case the electron and the neutrino are created at the same time as the neutron converts into a proton.) This both inspired and became the prototype for much of the subsequent exploration of the quantum character of the fundamental forces in nature.

Moreover, it didn't just make postdictions about nature. It made predictions precisely because a single mathematical form for the interaction that caused neutron decay could also predict a host of other phenomena, which were later observed.

Even more important, this interaction, with precisely the same strength, governs similar decays of other particles in nature. For example, in 1936 Carl Anderson, the discoverer of the positron, discovered another new particle in cosmic rays—the first of what would be so many that particle physicists would wonder whether the progression would ever end. When informed of this discovery, the atomic physicist and later Nobel laureate I. I. Rabi is said to have exclaimed, "Who ordered that?"

We now know that this particle, called the muon and characterized by the Greek letter μ, is essentially an exact copy of the electron, only about two hundred times heavier. Because it is heavier, it can decay, emitting an electron and a neutrino in an interaction that looks identical to neutron decay, except the muon converts into another type of neutrino (called the muon neutrino) instead of a proton. Remarkably, if we use the same Fermi constant for the strength of this interaction, we derive exactly the right lifetime for the muon.

Clearly a new fundamental force is at work here, universal in nature, with some similarities to electromagnetism, and some important differences. First, the interaction is much weaker. Second, unlike electromagnetism, the interaction appears to operate over only a small range—in Fermi's model at a single point. Neutrons don't turn into protons in one place and cause electrons to turn into neutrinos somewhere else, whereas the interaction between electrons and photons allows electrons to exchange virtual photons and be repelled by each other even at a great distance. Third, the interaction changes one type of particle into another. Electromagnetism involves the creation and absorption of photons—the quanta of light—but the charged particles that interact with them preserve their identity before and after the interaction. Gravity too is long-range, and when a ball falls toward the Earth, it remains a ball. But the weak interaction causes neutrons to decay into protons, muons into neutrinos, and so on.

Clearly something about the weak interaction is different, but you may wonder if it is worth worrying about. Neutron decay is interesting, but happily the properties of nuclei protect us from it so that stable atoms can exist. Thus it seems to have little impact on everyday lives. Unlike gravity and electromagnetism, we don't sense it. If the weak interaction were of little other importance, then its anomalous nature could be easily overlooked.

However, the weak interaction, at least as much as gravity and electromagnetism, is directly responsible for our existence. In 1939, Hans Bethe, who would soon help lead the effort to build the atomic bomb, realized that the interactions that broke apart heavy nuclei as the source of the explosive power of the bomb could, under different circumstances, be utilized to build larger nuclei from smaller nuclei. This could release even more energy than was released in the A-bomb.

Up until that time the energy source of the Sun was a mystery. It was well established that the temperature in the solar core could not exceed a few tens of millions of degrees—which may seem extreme, but the energies available to the colliding nuclei at those temperatures had already been achieved in the lab. Moreover, the Sun could not involve simple burning, like a candle.

It had been established as early as the eighteenth century that an object with the mass of the Sun could only burn with its observed brightness for perhaps ten thousand years if it were just something like a burning lump of coal. While that meshed nicely with Bishop Ussher's estimates for the age of the universe as inferred from the Bible's tale of creation, geologists and biologists had already established by the mid-nineteenth century that Earth itself was far older. With no apparent new energy source, the longevity and brightness of the Sun was inexplicable.

Enter Hans Bethe. Another of the incredibly talented and prolific theoretical physicists coming out of Germany in the first half of the twentieth century, Bethe was also another doctoral student of Arnold Sommerfeld's and also went on to win the Nobel Prize. Bethe began his career in chemistry because the introductory physics instruction at his university was poor—a common problem. (I also dropped physics in my first year for the same reason, but happily the physics department at my university let me take a more advanced course the following year.) Bethe switched to physics before moving on to graduate studies and emigrated to the United States to escape the Nazis.

A consummate physicist, Bethe could work through detailed calculations to solve a wide variety of problems on the blackboard, beginning at the upper left of the board and ending at the lower right with almost no erasures. Bethe strongly influenced Richard Feynman, who used to marvel at Bethe's patient methodological approach to problems. Feynman himself often jumped from the beginning of a problem to the end and worked out the steps in between afterward. Bethe's solid technical prowess and Feynman's brilliant insights combined well when they both worked at Los Alamos on the atomic bomb. They would go down the hallway with Feynman loudly countering the patient but persistent Bethe, and their colleagues labeled them "the Battleship and the Torpedo Boat."

Bethe was legendary when I was a young physicist because even into his nineties he was still writing important physics articles. He was also happy to talk to anyone about physics. When I gave a visiting lecture at Cornell—where Bethe spent most of his professional career—I felt immensely honored when he walked into my office to ask me questions and then listened intently to me, as if I actually had something to offer him.

He was also physically robust. A physicist friend of mine told me of a time he too visited Cornell. One weekend he decided to be ambitious and climb one of the many steep hiking trails near the campus. He was proud of himself for huffing and puffing his way almost to the top until he spied Bethe, then in his late eighties, happily making his way down the trail from the summit.

While I always liked and admired Bethe, in researching material for this book I found two additional happy personal connections that were satisfying enough for me to relate them here. First, I found out that I am in a sense his intellectual grandson, as my undergraduate physics honors thesis adviser, M. K. Sundaresan, was one of his doctoral students. Second, I discovered that Bethe, who had little patience for grand claims made of fundamental results that were carried out without any real motivation or evidence, once wrote a hoax paper while a postdoc poking fun at a paper he deemed ridiculous by the famous physicist Sir Arthur Stanley Eddington. Eddington claimed to "derive" a fundamental constant of electromagnetism using some fundamental principles, but Bethe correctly viewed the claim as nothing other than misguided numerology. Learning this made me feel better about a hoax paper I wrote when I was an assistant professor at Yale, responding to what I thought was an inappropriate paper, published in a distinguished physics journal, that claimed to discover a new force in nature (which indeed later turned out to be false). At the time that Bethe wrote his paper, the physics world took itself a little more seriously, and Bethe and his colleagues were forced to issue an apology. By the time I wrote mine, the only negative reaction I got was from my department chair, who was worried that the *Physical Review* might actually publish my article.

When he was in his early thirties, Bethe had already established himself as a master physicist with his name attached to a host of results, from the Bethe formula, describing the passage of charged particles through matter, to the Bethe ansatz, a method to obtain exact solutions for certain quantum problems in many-body physics. A series of reviews he cowrote on the state of the nascent field of nuclear physics in 1936 remained authoritative for some time and became known as Bethe's Bible. (Unlike the conventional Bible, it made testable predictions, and it was eventually replaced as scientific progress was made.)

In 1938, Bethe was induced to attend a conference on "stellar energy generation," though at that time astrophysics was not his chief interest. By the end of the meeting, he had worked out the nuclear processes by which four individual protons (the nuclei of hydrogen atoms) eventually "fuse"—as a result of Fermi's weak interaction—to form the nucleus of helium, containing two protons and two neutrons. This fusion releases about a million times more energy per atom than is released when coal burns. This allows the Sun to last a million times longer than previous estimates would have permitted, or about 10 billion years instead of ten thousand years. Bethe later showed that other nuclear reactions help power the Sun, including a set that converts carbon to nitrogen and oxygen—the so-called CNO cycle.

The secret of the Sun—the ultimate birth of light in our solar system—had been unveiled. Bethe won the Nobel Prize in 1967, and almost forty years after that, experiments on neutrinos coming from the Sun confirmed Bethe's predictions. Neutrinos were the key experimental observable that allowed such confirmation. This is because the whole chain begins with a reaction in which two protons collide, and via the weak interaction one of them converts into a neutron, allowing the two to fuse into the nucleus of heavy hydrogen, called deuterium, and release a neutrino and a positron. The positron later interacts in the Sun, but neutrinos, which interact only via the weak interaction, travel right out of the Sun, to Earth and beyond.

Every second of every day, more than 400,000 billion of these neutrinos are passing through your body. Their interaction strength is so weak that they could traverse on average through ten thousand light-years of solid lead before interacting, so most of them travel right through you, and Earth, without anyone's noticing. But if not for the weak interaction, they would not be produced, the Sun wouldn't shine, and none of us would be here to care.

So the weak interaction, although extremely weak, nevertheless is largely responsible for our existence. Which is one of the reasons why, when the Fermi interaction, developed to characterize it, and the neutrinos first predicted by it, turned out to both defy common sense, physicists had to stand up and take notice. And they were driven to change our notions of reality itself.

Part Two

EXODUS

Chapter 11

DESPERATE TIMES
AND DESPERATE MEASURES

To every thing there is a season, and a time for
every purpose.

—ECCLESIASTES 3:1

The rapid succession of events during the 1930s, from the discovery of the neutron to probing the nature of neutron decay, as well as the discovery of the neutrino and the consequent discovery of a new and universal short-range weak force in nature, left physicists more confused than inspired. The brilliant march that had led to the unification of electricity and magnetism, and the unification of quantum mechanics and relativity, had been built on exploring the nature of light. Yet it wasn't clear how the elegant theoretical edifice of quantum electrodynamics could guide considerations of a new force. The weak interaction is far removed from direct human experience and involves new and exotic elementary particles and nuclear transmutations reminiscent of alchemy but, unlike alchemy, testable and reproducible.

The fundamental confusion lay with the nature of the atomic nucleus itself and the question of what held it together. The discovery of the neutron helped resolve the paradox that had earlier seemed to require electrons to be confined in the nucleus to counter the charge of additional protons necessary to produce correct nuclear masses, but the observation of beta decay—which resulted in electrons emerging from nuclei—didn't help matters.

The realization that in beta decay neutrons became protons in the nucleus clarified matters, but then another question naturally arose: Could this transformation somehow explain the strong binding that held protons and neutrons together inside nuclei?

In spite of the obvious differences between the weak forces and quantum theory of electromagnetism, QED, the remarkable success of QED in describing the behavior of atoms and the interactions of electrons with light colored physicists' thinking about the new weak force as well. The mathematical symmetries associated with QED worked beautifully to ensure that otherwise worrisome infinities in the calculations arising from the exchange of virtual particles vanished when making predictions of physical quantities. Would something similar work to understand the force binding protons and neutrons in nuclei?

Specifically, if the electromagnetic force was due to the exchange of particles, then it was reasonable to think that the force that held together the nucleus might also be due to the exchange of particles. Werner Heisenberg proposed this idea in 1932 around the time the neutron was discovered. If neutrons and protons could convert into each other, with the proton absorbing an electron to become a neutron, then maybe the exchange of electrons between them might somehow produce a binding force?

A number of well-known problems marred this picture, however. First was the problem of "spin." If one assumed, as Heisenberg did, that the neutron was essentially made up of a proton and an electron bound together, and since both were spin ½ particles, then adding them together in the neutron, it couldn't have spin ½ as well, since ½ + ½ can't equal ½. Heisenberg argued, in desperation, because those were desperate times when it seemed all the conventional rules were breaking down, that the "electron" that was transferred between neutrons and protons, and which bound them together in the nucleus, was somehow different from a free electron and had no spin at all.

In retrospect, this picture has another problem. Heisenberg was motivated to consider electrons binding together neutrons and protons because he was thinking about hydrogen molecules. In hydrogen, two protons are bound together by sharing electrons that orbit them. The problem with using a similar explanation for nuclear binding is one of scale. How could neutrons and protons exchange electrons and be bound together so tightly that their average distance apart is more than one hundred thousand times smaller than the size of hydrogen molecules?

Here is another way of thinking about this problem that will be useful to return to later. Recall that electromagnetism is a long-range force. Two electrons on opposite sides of the galaxy experience a repulsion—albeit extremely small—due to the exchange of virtual photons. The quantum theory of electromagnetism makes this possible. Photons are massless, and virtual photons can travel arbitrarily far, carrying arbitrarily small amounts of energy, before they are absorbed again—without violating the Heisenberg uncertainty principle. If the photons were massive, then this would not be possible.

Now if a force between neutrons and protons in nuclei arose due to the absorption and emission of virtual electrons, say, then the force would be short-range because the electrons are massive. How short-range? Well, it works out to be about one hundred times the size of typical nuclei. So, exchanging electrons doesn't work to produce nuclear-scale forces. As I say, those were desperate times.

Heisenberg's desperate idea about a strange spinless version of the electron was not lost on a young Japanese physicist, the shy twenty-eight-year-old Hideki Yukawa. Working in 1935 when Japan was just beginning to emerge from centuries of isolation, and just before its imperial designs ignited the war in the Pacific, Yukawa published the first original work in physics to be published by a physicist educated entirely in Japan. No one took notice of the paper for at least two years, yet fourteen years later he won the Nobel Prize for this work, which *had* by then become noticed, but for the wrong reasons.

Einstein's visit to Japan in 1922 had cemented Yukawa's growing interest in physics. When Yukawa was still in high school and searching for material to help him pass examinations in a second foreign language, he found Max Planck's *Introduction to Theoretical Physics* in German. He rejoiced in reading both the German and the physics and was aided by his classmate Sin-Itiro Tomonaga, a talented physicist who was his colleague both in high school and later at Kyoto University. Tomonaga was so talented that he would later share the 1965 Nobel Prize with Richard Feynman and Julian Schwinger for demonstrating the mathematical consistency of quantum electrodynamics.

That Yukawa, who had been a student in Japan at a time when many of his instructors did not yet fully understand the emerging field of quantum mechanics, came upon a possible solution to the nuclear-force problem that had been overlooked by Heisenberg, Pauli, and even Fermi was remarkable. I suspect that part of the problem was a phenomenon that has occurred several times in the twentieth century and perhaps before, and perhaps after. When the paradoxes and complexities associated with some physical process begin to seem overwhelming, it is tempting to assume that some new revolution, similar to relativity or quantum mechanics, will require such a dramatic shift in thinking that it doesn't make sense to push forward with existing techniques.

Fermi, unlike Heisenberg or Pauli, was not looking for a wholesale revolution. He was willing to propose, as he called it, a "tentative theory" of neutron decay that got rid of electrons in the nucleus by allowing them to be spontaneously created during beta decay. He proposed a model that worked, which he knew was just a model and not a complete theory, but it did allow one to do calculations and make predictions. That was the essence of Fermi's practical style.

Yukawa had followed these developments, translated Heisenberg's paper on nuclei along with an introduction, and published it in Japan, so the problems of Heisenberg's proposal were already clear to him. Then in 1934 Yukawa read Fermi's theory of neutron decay, which catalyzed a new idea in Yukawa's mind. Perhaps the nuclear force binding protons and neutrons was due not to the exchange of virtual electrons between them, but to the exchange of both the electron and the neutrino that were created when neutrons changed to protons.

Another problem immediately arose, however. Neutron decay is a result of what would become known as the weak interaction, and the force responsible for it is weak. Plugging in values for the possible force that might result between protons and neutrons by the exchange of an electron-neutrino pair made it clear that this force would be far too weak to bind them.

Yukawa then allowed himself to do what none of the others had done. He questioned why the nuclear force, if it, like QED, results from the exchange of virtual particles, had to be due to the exchange of one or more of the particles already known or assumed to exist. Remembering how loath physicists such as Dirac and Pauli had been to propose new particles, even when they were correct, you can perhaps appreciate how radical Yukawa's idea was. As Yukawa later described it:

At this period the atomic nucleus was inconsistency itself, quite inexplicable. And why?—because our concept of elementary particle was too narrow. There was no such word in Japanese and we used the English word—it meant proton and electron. From somewhere had come a divine message forbidding us to think about any other particle. To think outside of these limits (except for the photon) was to be arrogant, not to fear the wrath of the gods. It was because the concept that matter continues forever had been a tradition since the times of Democritus and Epicurus. To think about creation of particles other than photons was suspect, and there was a strong inhibition of such thoughts that was almost unconscious.

One of my good physics friends has said that the only time he was able to do complicated calculations was after the birth of each of his children, when he couldn't sleep anyway, so he stayed up and worked. Thus in October of 1934, just after the birth of his second child and unable to sleep, Yukawa realized that if the range of the strong nuclear force was to be restricted to the size of a nucleus, then any exchanged particle must be far more massive than the electron. The next morning he estimated the mass to be two hundred times the electron mass. It would have to carry an electric charge if it was to be exchanged between neutrons and protons, and it could have no spin, so as not to change the proton's or neutron's spin when it was absorbed or emitted.

What has all this concern over strong nuclear forces to do with neutron decay, the subject that started this chapter and ended the last? you may ask. In the 1930s, just as it went against the grain to imagine new particles, so too inventing new forces seemed unnecessary at best and heretical at worst. Physicists were convinced that all the processes that occurred in the nucleus, strong or weak, must be connected.

Yukawa envisaged a clever way to do this, connecting ideas of both Fermi and Heisenberg, and also generalizing ideas from the successful quantum theory of electromagnetism. If instead of emitting a photon, neutrons in the nucleus emitted a new, heavy, spinless charged particle, which Yukawa originally called a mesotron—until Heisenberg corrected Yukawa's Greek and the name was shortened to meson—then that particle could be absorbed by protons in the nucleus, producing a force of attraction whose magnitude Yukawa was able to calculate using equations that were extrapolated from, you guessed it, electromagnetism.

The analogy with electromagnetism could not be exact, however, because the meson is massive and the photon is massless. Yukawa took the attitude that Fermi might have, if he had thought of it. Yes, the theory wasn't complete, but Yukawa was willing to ignore the other aspects of electromagnetism that this theory couldn't reproduce. Damn the torpedoes, full speed ahead.

Yukawa ingeniously—and ultimately incorrectly—connected this strong force to observed neutron decay by suggesting that mesons might not always simply be exchanged between neutrons and protons in the nucleus. A small fraction of the mesons emitted by neutrons might decay en route into an electron and neutrino before they could be re-absorbed, causing neutron decay. In this case, the neutron decay would not be described by something like the figure below and on the left, where the decay and the emission of the other particles all occur at a single point. It would appear like the figure on the right, where the decay gets spread out and a new particle, shown by the dashed line (which represents Yukawa's meson), travels a short distance after emission before decaying into the electron and neutrino. With the new intermediate particle, the weak interaction mediating neutron decay begins to look more like the electromagnetic interaction between charged particles:

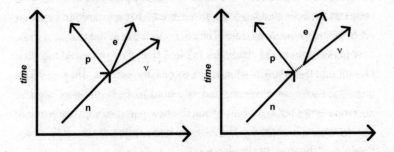

Yukawa had proposed a new intermediate particle, a heavy meson, which made neutron decay look similar to the earlier picture of photon exchange in electromagnetism—which had motivated his thinking in the first place—but with significant differences. In this case the intermediate particle was both massive and electrically charged, and also unlike the photon it had no spin angular momentum.

Nevertheless, Yukawa was able to show that for a heavy meson his theory would be indistinguishable from Fermi's point interaction describing neutron decay—at least for predicting the details of neutron decay. In addition, Yukawa's theory offered the possibility of reducing all of the strange properties of the nucleus—from beta decay of neutrons inside the nucleus to the strength of the interaction binding together protons and neutrons—to merely understanding the properties of a single new interaction, due to the exchange of a new particle, his meson.

However, if this new heavy meson existed, where was it? Why hadn't it yet been seen in cosmic rays? Because of this, and also because Yukawa was an unknown entity working in a location far from all the action, no real attention was paid to his proposal to explain both the strong interaction between nucleons and the weaker one that appeared to be responsible for neutron decay. Nevertheless, his proposal, unlike those of Heisenberg and others (including Fermi), was simpler and made more sense.

All of this changed in 1936, less than two years after Yukawa's prediction, when Carl Anderson, the discoverer of the positron, together with collaborator Seth Neddermeyer, discovered what appeared to be a new set of particles in cosmic rays. The characteristics of the tracks of these new particles in cloud chambers implied that they produced too little radiation in traversing matter to be protons or electrons. They were also more massive than electrons and appeared to be sometimes negative and sometimes positive. Before long the new particles were determined to have a mass in the range that Yukawa had predicted—about two hundred times the mass of the electron.

It is remarkable how quickly the rest of the world caught on. Yukawa published a short note to point out that his theory predicted just such particles. Within weeks the major physicists in Europe began exploring his model and incorporating his ideas in their work. In 1938, in the last major conference before the Second World War interrupted essentially all international collaborations in science, of the eight main speakers, three dealt with Yukawa's theory—citing a name they would have been unfamiliar with a year or two before.

While much of the rest of the physics world celebrated the apparent discovery of Yukawa's meson, this discovery was not without its own problems. In 1940 the decay of a meson to an electron, predicted by Yukawa, was observed in cosmic-ray tracks. However, over the years 1943 to 1947 it became clear that the particles Anderson and Neddermeyer had discovered interacted much more weakly with nuclei than Yukawa's particle should have.

Something was wrong.

Three of Yukawa's Japanese colleagues suggested that mesons were of two different sorts, and that a Yukawa-type meson might decay into yet another, different and more weakly interacting meson. But their articles were in Japanese and didn't appear in English until after the war, by which time a similar proposal had been made by the US physicist Robert Marshak.

This delay proved fortuitous. New techniques were being developed to observe the tracks of cosmic rays in photographic emulsions, and a series of brave researchers dragged their equipment up to high elevations to search for possible new signals. Many cosmic rays interact and disappear before reaching sea level, so this group and others interested in exploring this wondrous new source of particles coming from the heavens had no choice but to seek higher elevations. Here cosmic rays would have traversed less distance in the atmosphere and might be more easily detected.

The former Italian mountain guide turned physicist Giuseppe Occhialini had been invited from Brazil to join a British team working on the A-bomb during the war. As a foreign national, he couldn't work on the project, so instead he joined the cosmic-ray physics group at Bristol. Occhialini's mountain training proved useful as he dragged photographic emulsions up to the Pic du Midi at twenty-eight hundred meters in France. Today you can travel to the observatory on top of this peak by cable car, and it is a terrifyingly exciting ride. But in 1946 Occhialini had to climb to the top, risking his health in the effort to discover signals of exotic new physics.

And he and his team did discover exotic new physics. As Cecil Powell, Occhialini's collaborator at Bristol (and future Nobel laureate, while Occhialini, who had done the climbing, did without), put it, they saw "a whole new world. It was as if, suddenly, we had broken into a walled orchard, where protected trees flourished and all kinds of exotic fruits had ripened in great profusion."

Less poetically, perhaps, what they discovered were two examples in which an initial meson stopped in the emulsion and gave rise to a second meson, just as had been suggested by the theorists. Many more events were observed with emulsions taken to an elevation almost twice as high as Pic du Midi. In October of 1947, in the journal *Nature*, Powell, Occhialini, and Powell's student Cesare Lattes published a paper in which they named the initial meson the pion—which seemed to interact with the nuclear strength appropriate to Yukawa's meson—and the subsequent meson the muon.

It seemed at long last that Yukawa's meson had been discovered. As for its "partner" the muon, which had been confused with Yukawa's meson, it was nothing of the sort. Not spinless, it instead had the same spin as the electron and the proton. And its interactions with matter were nowhere near strong enough to play a role in nuclear binding. The muon turned out to be simply a heavy, if unstable, copy of the electron, which is what motivated Rabi's question "Who ordered that?"

So, the particle that made Yukawa famous wasn't the particle he predicted after all. His idea became famous because the original experimental result had been misinterpreted. Fortunately, the Nobel committee waited until the 1947 discovery of the pion before awarding Yukawa their prize in 1949.

But, given the track record of errors and mislabeling, it is natural to wonder if the pion was in fact the particle Yukawa had predicted. The answer is both yes and no. Exchange of charged pions between protons and neutrons is indeed one accurate way of trying to estimate the strong nuclear force holding nuclei together. But in addition to charged pions—

the mesons that Yukawa had predicted—there are neutral pions as well. Who ordered those?

Moreover, the theory that Yukawa wrote down to describe the strong force, like Fermi's theory to describe neutron decay, was not fully mathematically consistent, as Yukawa had conceded when he proposed it. There *was*, at the time, no correct relativistic theory involving the exchange of massive particles. Something was still amiss, and a series of surprising experimental discoveries, combined with prescient theoretical ideas that were unfortunately applied to the wrong theories, helped lead to more than a decade of confusion before the fog lifted and light appeared at the end of the tunnel. Or perhaps at the mouth of the cave.

Chapter 12

MARCH OF THE TITANS

The wolf also shall dwell with the lamb, and the
leopard shall lie down with the kid.

—ISAIAH 11:6

The relationship between theoretical insight and experimental discovery is one of the most interesting aspects of the progress of science. Physics is at its heart, like all of science, an empirical discipline. Yet at times brief bursts of theoretical insight change everything. Certainly Einstein's insights into space and time in the first two decades of the twentieth century are good examples, and the remarkable theoretical progress associated with the development of quantum mechanics by Schrödinger, Heisenberg, Pauli, Dirac, and others in the 1920s is another.

Less heralded is another period, from 1954 to 1974, which, while not as revolutionary, will, when sufficient time has passed, be regarded as one of the most fruitful and productive theoretical physics eras in the twentieth century. These two decades took us, not without turmoil, from chaos to order, from confusion to confidence, and from ugliness to beauty. It's a wild ride, with a few detours that might seem to come from left field, but bear with me. If you find it a tad uncomfortable, then recall what I said in the introduction about science and comfort. By putting yourself in the frame of mind of those involved in the quest, whose frustration eventually led to insights, the significance of the insights can be truly appreciated.

This tumultuous period followed one in which experimental bombshells had produced widespread confusion, making nature "curiouser and curiouser," as Lewis Carroll might have put it. The discoveries of the positron and quickly thereafter the neutron were just the beginning. Neutron decay, nuclear reactions, muons, pions, and a host of new elementary particles that followed made it appear as if fundamental physics was hopelessly complicated. The simple picture of a universe in which electromagnetism and gravity alone governed the interactions of matter made from protons and electrons disappeared into the dustbin of history. Some physicists at the time, like some on the political right today, yearned for the (often misremembered) simplicity of the good old days.

This newfound complexity drove some, by the 1960s, to imagine that *nothing* was fundamental. In a Zen-like picture, they imagined that all elementary particles were made from all other elementary particles, and that even the notion of fundamental forces might be an illusion.

Nevertheless, percolating in the background were theoretical ideas that would draw back the dark curtains of ignorance and confusion, revealing an underlying structure to nature that is as remarkable as it is strangely simple, and one in which light would once again play a key role.

It all began with two theoretical developments, one profound and unheralded and another relatively straightforward but brilliant and immediately feted. Remarkably, the same man was involved in both.

Born in 1922 to a mathematician father, Chen-Ning Yang was educated in China, moving in 1938 from Beijing to Kunming to avoid the Japanese invasion of China. He graduated four years later from National Southwestern Associated University and remained there for another two years. There he met another student who had been forced to re-locate to Kunming, Tsung-Dao Lee. While they only had a marginal acquaintance with the United States, in 1946 both of them received scholarships set up by the US government, with funds received from China to allow talented Chinese students to pursue graduate study in America. Yang had a master's degree and therefore had greater freedom to pursue a PhD, and went with Fermi from Columbia to the University of Chicago. Lee had less choice, as he did not have a master's degree, but the only US university where he could work directly toward a PhD was also the University of Chicago. Yang did his PhD under the supervision of Edward Teller and worked directly with Fermi as his assistant for only a year after graduation, while Lee did his PhD with Fermi directly.

During the 1940s, the University of Chicago was one of the greatest centers of theoretical and experimental physics in the country, and its graduate students benefited from their exposure to a remarkable set of scientists—not only Fermi and Teller, but others including the brilliant but unassuming astrophysicist Subrahmanyan Chandrasekhar. When he was nineteen, Chandra, as he was often called by colleagues, had proved that stars greater than 1.4 times the mass of the Sun must col-lapse catastrophically at the end of their nuclear-burning lifetime, either through what is now known to be a supernova explosion, or directly in what is now known as a black hole. While his theory was ridiculed at the time, he was awarded the Nobel Prize for that work fifty-three years later.

Chandra was not just a brilliant scientist but, like Fermi, a dedicated teacher. Even though he was pursuing research at the Yerkes Observa-tory in Wisconsin, he drove one hundred miles round-trip each week to teach a class to just two registered students, Lee and Yang. Ultimately, the entire class, professor included, became Nobel laureates, which is probably unique in the history of science.

Yang moved to the venerable Institute for Advanced Study in Prince-ton in 1949, where he nurtured his budding collaboration with Lee on a variety of topics. In 1952 Yang was made a permanent member of the institute, while Lee moved in 1953 to nearby Columbia in New York City, where he remained for the rest of his career.

Each of these men made major contributions to physics in a variety of areas, but the collaboration that made them famous began with a strange experimental result, again coming from cosmic-ray observations.

In the same year that Yang moved from Chicago to the IAS, Cecil Powell, the discoverer of the pion, discovered a new particle in cosmic rays, which he called the tau meson. This particle was observed to decay into three pions. Another new particle was discovered shortly thereafter, called the theta meson, which decayed into two pions. Surprisingly, this new particle turned out to have precisely the same mass and lifetime as that tau meson.

This might not seem that strange. Might they be the same particle, simply observed to decay in two different ways? Remember that in quantum mechanics, anything that is not forbidden can happen, and as long as the new particle was heavy enough to decay into either two or three pions—and the weak force allowed such decays—both should occur.

But, if it were sensible, the weak force shouldn't have allowed both decays.

Think for a moment about your hands. Your left hand differs from your right hand. No simple physical process, short of entering through the looking glass, can convert one into the other. No series of movements, up or down, turning around, or jumping up and down, can turn one into the other.

The forces that govern our experience, electromagnetism and gravity, are blind to the distinction between left and right. No process moderated by either force can turn something such as your right hand into its mirror image. I cannot turn your right hand into your left hand merely by shining light on it, for example.

Put another way, if I shine a light on your right hand and look at it from a distance, the intensity of reflected light will be the same as it would be if I did the same thing to your left hand. The light doesn't care about left or right when it is reflecting off something.

Our definition of left and right is imposed by human convention. Tomorrow we could decide that left is right and vice versa, and nothing would change except our labels. As I write this on an airplane, flying economy class, the person to my right may be quite different from the person to my left, but again that is just an accident of my circumstances. I don't expect that the laws governing the flight of this plane are different for the right wing than for the left wing.

Think about this in the subatomic world. Recall that Enrico Fermi found that, given the rules of quantum mechanics, the mathematical behavior of groups or pairs of elementary particles depends on whether they have spin ½, i.e., are fermions. The behavior of groups of fermions is quite different from the behavior of particles such as photons, which have a spin value of 1 (or any integer value of spin angular momentum, i.e., 0, 1, 2, 3, etc.). The mathematical "wave function" that describes a pair of fermions, for example, is "antisymmetric," while one describing a pair of photons is "symmetric." This means that if one interchanges one particle with another, the wave function describing fermions changes sign. But for particles such as photons, the wave function remains the same under such an interchange.

Interchanging two particles is the same as reflecting them in the mirror. The one on the left now becomes the one on the right. Thus an intimate connection exists between such exchanges and what physicists call parity, which is the overall property of a system under reflection (i.e., interchanging left and right).

If an elementary particle decays into two other particles, the wave function describing the "parity" of the final state (i.e., whether the wave function changes sign or not under left-right interchange of the particles) allows us then to assign a quantity we can call parity to the initial particle. In quantum mechanics if the force that governs the decay is blind to left and right, then the decay will not change the parity of the quantum state of the system.

If the wave function of the system is antisymmetric under interchange of the particles after the decay, then the system has "negative" parity. In this case the wave function describing the initial quantum state of the decaying particle must also have negative parity (i.e., it would change sign if left and right were interchanged).

Now, pions, the particles discovered by Powell and hypothesized by Yukawa, have negative parity, so that the wave function that describes the quantum state of their mirror image would change sign compared to the original wave function. The distinction between positive and negative parity is kind of like considering first a nice spherical ball, which looks identical when reflected in the mirror, and hence has positive parity:

Versus, say, your hand, which changes character (from left to right) when reflected in a mirror and could therefore be said to have negative parity:

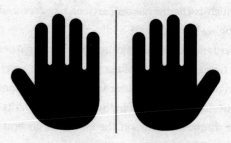

These somewhat abstract considerations made the observed data on the decays of the new particles that Powell discovered perplexing. Because a pion has negative parity, two pions would have positive parity, since $(-1)^2 = 1$. A system of three pions, however, would, by the same consideration, have negative parity, since $(-1)^3 = -1$. Therefore if parity doesn't change when a particle decays, a single original particle cannot decay into two different final states of different parity.

If the force responsible for the decay behaved like all the other known forces at the time, such as electromagnetism or gravity, it would be blind to parity (it would not distinguish between right and left), so it shouldn't change the original parity of the system after the decay, just as shining a light on your right hand will not cause it to look like your left hand.

Since it seemed impossible for a single type of particle to decay sometimes into two, and sometimes into three, pions, the solution seemed simple. There must be two different new elementary particles, with opposite parity properties. Powell dubbed these the tau particle and theta particle—one of which could decay into two pions, and one into three pions.

Observations suggested that the two particles had precisely the same masses and lifetimes, which was a bit strange, but Lee and Yang proposed that this might be a general property for various elementary particles, which they suggested might come in pairs with opposite parity. They called this idea "parity doubling."

Such was the situation in the spring of 1956 when the International Conference on High Energy Physics, held every year at the University of Rochester, took place. In 1956, the entire community of physicists interested in particle and nuclear physics could fit in a single university lecture hall, and these physicists, including all the major players, tended to gather at this annual meeting. Richard Feynman was sharing a room at the meeting with Marty Block. Being an experimentalist, Block was not as burdened by the possible heresy inherent in the suggestion that some force in nature was not blind to the distinction between left and right, and he asked Feynman if possibly the weak interaction governing the decays Powell observed might distinguish left from right. This would allow a single particle to decay to states of differing parity—meaning the tau and theta could both be the same particle.

Block didn't have the temerity to raise this question in the public session, but Feynman did, even though he privately thought this was extremely unlikely. Yang replied that he and Lee had thought about this, but so far nothing had come of the idea. Eugene Wigner, who would later win a Nobel Prize for elucidating the importance of such things as parity in atomic and nuclear physics, was also present, and he too raised the same question about the weak interaction.

But to the victor go the spoils, and speculating about the possible violation of parity by a new force in nature that might distinguish left from right was different from demonstrating it. A month later Lee and Yang were at a café in New York, and they decided to examine all known experiments involving the weak interaction to see if any of them could dispel the possibility of parity violation. To their great surprise, they realized that not a single one definitively resolved the issue. As Yang later said, "The fact that parity conservation in the weak interaction was believed for so long without experimental support was very startling. But what was more startling was the prospect that a space-time symmetry law which the physicists have learned so well may be violated. This prospect did not appeal to us."

To their credit, Lee and Yang proposed a variety of experiments that *could* test the possibility that the weak interaction distinguished right from left. They suggested considering the beta decay of a neutron in the nucleus of cobalt-60. Because this radioactive nucleus has nonzero spin angular momentum—i.e., it behaves as if it is spinning—it also acts like a little magnet. In an external magnetic field the nuclei will line up in the direction of the field. If the electron emitted when a neutron in the nucleus decays preferentially ends up in one hemisphere instead of another, this would be a sign of parity violation, because in the mirror the electrons would end up in the opposite hemisphere.

If this was true, then at a fundamental level, nature would be able to distinguish right from left. The human-created distinctions between them (i.e., sinister versus good) would not then be totally artificial. Thus the world in a mirror could be distinguished from the real world, or, as Richard Feynman poetically put it later, we could use this experiment to send a message to tell a Martian what direction is "left"—say, the hemisphere where more electrons were observed to emerge—without drawing a picture.

At the time, this was viewed as such a long shot that many in the physics community were amused, but no one ran out to perform the experiment. No one, that is, except Lee's colleague at Columbia the experimentalist Chien-Shiung Wu, known as Madame Wu.

Even as we bemoan today the paucity of female physicists trained at American institutions, the situation was much worse in 1956. After all, women weren't even admitted as undergraduates at Ivy League institutions until the late 1960s. Almost thirty years after Wu arrived from China to study at Berkeley in 1936, she noted in a *Newsweek* article about her, "It is shameful that there are so few women in science. . . . In China there are many, many women in physics. There is a misconception in America that women scientists are all dowdy spinsters. This is the fault of men. In Chinese society, a woman is valued for what she is, and men encourage her to accomplishments—yet she remains eternally feminine."

Be that as it may, Wu was an expert in neutron decay and became intrigued by the tantalizing possibility of searching for parity violation in the weak interaction after learning of it from her friends Lee and Yang. She canceled a European vacation with her husband and embarked on an experiment in June, one month after Lee and Yang had first thought of the problem, and by October of that year—the same month Lee and Yang's paper appeared in print—she and several colleagues had assembled the apparatus necessary to do the experiment. Two days after Christmas of that year they had a result.

In modern times particle physics experiments might take decades from design to completion, but that was not the case in the 1950s. It was also a time when physicists apparently didn't bother to take holidays. Despite its being the yuletide, the Friday "Chinese Lunches" organized by Lee continued, and the first Friday after New Year's Day Lee announced that Wu's group had discovered that not only was parity violated, but it was violated by the maximum amount possible in the experiment. The result was so surprising that Wu's group continued their work to ensure they weren't being fooled by an experimental glitch.

Meanwhile, Leon Lederman and colleagues Dick Garwin and Marcel Weinrich, also at Columbia, realized that they could check the result in their experiments on pion and muon decays at Columbia's cyclotron. Within a week, both groups, as well as Jerry Friedman and Val Telegdi in Chicago, independently confirmed the result with high confidence, and by mid-January 1957 they submitted their papers to the *Physical Review*. They changed our picture of the world forever.

Columbia University called what was probably the first press conference ever announcing a scientific result. Feynman lost a $50 bet, but Wolfgang Pauli was luckier. He had written a letter from Zurich on January 15 to Victor Weisskopf at MIT betting that Wu's experiment would not show parity violation, not knowing that the experiment already had. Pauli exclaimed in the letter, "I refuse to believe that God is a weak lefthander," demonstrating an interesting appreciation for baseball as well. Weisskopf, who by then knew of the actual result, was too kind to take the bet.

Upon hearing the news, Pauli later wrote, "Now that the first shock is over, I begin to collect myself." It really was a shock. The idea that one of the fundamental forces in nature distinguished between right and left flew in the face of common sense, as well as of much of the basis of modern physics as it was understood then.

The shock was so great that, for one of the few times in the history of the Nobel Prizes, Nobel's will was actually carried out properly. His will stipulates that the prize should go to the person or persons in each field whose work *that year* was the most important. In October of 1957, almost exactly a year from the publication of Lee and Yang's paper, and only ten months after Wu and Lederman confirmed the notion, the thirty-year-old Lee and the baby-faced thirty-four-year-old Yang shared the Nobel Prize for their proposal. Sadly, Madame Wu, known as the Chinese "Madame Curie," had to be content with winning the inaugural Wolf Prize in Physics twenty years later.

Suddenly the weak interaction became more interesting, and also more confusing. Fermi's theory, which had sufficed up to that point, was roughly modeled after electromagnetism. We can think of the electromagnetism interaction as a force between two different electric currents, each corresponding to the two separate moving electrons that interact with each other. The weak interaction could be thought of in a somewhat similar way, if in one current a neutron, during the interaction, converts into a proton, and in the other current is an outgoing electron and neutrino.

There are two crucial differences, however. In Fermi's weak interaction the two different currents interact at a single point rather than at a distance, and the currents in the weak interaction allow particles to change from one type to another as they extend through space.

While electromagnetic interactions are the same in the mirror as they are in the real world, if parity is violated in the weak interaction, the "currents" involved would have to have a "handedness," as Pauli alluded, as for example a corkscrew or pair of scissors has, so that their mirror images will not be the same.

Parity violation in weak interactions would then be like the social rule that we always shake hands with our right hand. In a mirror world, people would always shake with their left hand. Thus, the real world differs from its mirror image. If the currents in the weak interaction had a handedness, then the weak interaction could distinguish right from left and in a mirror world would be different from the force in the world in which we live.

A great deal of work and confusion resulted as physicists tried to figure out precisely what types of new possible interaction could replace Fermi's simple current-to-current interaction, in which no apparent handedness could be attributed to the particles involved. Relativity allowed a variety of possible generalizations of Fermi's interaction, but the results of different experiments led to different, mutually exclusive mathematical forms for the interaction, so it appeared impossible that one universal weak interaction could explain all of them.

Around the time when the first experimental results on neutron and muon decay had come out suggesting that parity violation was as large as it could be, a young graduate student at the University of Rochester, George Sudarshan, began exploring the confused situation and came up with what eventually was the correct form of a universal interaction that could replace Fermi's form—something that also required that at least some of the experimental results at the time were wrong.

The rest of the story is a bit tragic. At the Rochester conference three months after the parity-violation discovery, and a year after Lee and Yang had presented their first thoughts on parity doubling, Sudarshan asked to present his results. But because he was a graduate student, he wasn't allowed. His supervisor, Robert Marshak, who had suggested the research problem to Sudarshan, was by then preoccupied with another problem in nuclear physics and chose to present a talk on that subject at the meeting. Another faculty member, who was asked to mention Sudarshan's work, also forgot. So all of the discussion at the meeting on the possible form of the weak interaction ended up leading nowhere.

Earlier, in 1947, Marshak had been the first to suggest that two different mesons were discovered in Cecil Powell's experiments—with one being the particle proposed by Yukawa, and the other being the particle now called a muon. Marshak was also the originator of the Rochester conferences and probably felt it would show favoritism to allow his own student to speak. In addition, since Sudarshan's idea required at least some of the experimental data to be wrong, Marshak may have decided it was premature to present it at the meeting.

That summer Marshak was working at the RAND Corporation in Los Angeles and invited Sudarshan and another student to join him. The two most renowned particle theorists in the world then, Feynman and Murray Gell-Mann, were at Caltech, and each had become obsessed with unraveling the form of the weak interaction.

Feynman had missed out on the discovery of parity violation by not following his own line of questioning, but had since realized that his work on quantum electrodynamics could shed light on the weak interaction. He desperately wanted to do this because he felt his work on QED was simply a bit of technical wizardry and far less noble than unearthing the form of the law governing another of the fundamental interactions in nature. But Feynman's proposal for the form of the weak interaction also appeared to disagree with experiments at the time.

Over the 1950s, Gell-Mann would produce many of the most important and lasting ideas in particle physics from that time. He was one of two physicists to propose that protons and neutrons were made of more fundamental particles, which he called quarks. He had his own reasons for thinking about parity and the weak interaction. Much of his success was based on focusing on new mathematical symmetries in nature, and he had used these ideas to come up with a new possible form for the weak interaction as well, but again his idea conflicted with experiment.

While they were in LA, Marshak arranged for Sudarshan to have lunch with Gell-Mann to talk about their ideas. They also met with an eminent experimentalist, Felix Boehm, whose experiments, he said, were now consistent with their ideas. Sudarshan and Marshak learned from Gell-Mann that his ideas were consistent with Sudarshan's proposal, but that at best Gell-Mann was planning to include the notion in one paragraph of a long general paper on the weak interaction.

Meanwhile, Marshak and Sudarshan prepared a paper on their idea, and Marshak decided to save it for a presentation at an international conference in Italy in the fall. Learning of the new experimental data from Boehm, Feynman decided—rather excitedly—that his ideas were correct and began to write a paper on the subject. Gell-Mann, who was competitive in the extreme, decided he too should write up a paper since Feynman was writing one. Eventually their department chairman convinced them they needed to write their paper together, which they did, and it became famous. Although the paper had an acknowledgment to Sudarshan and Marshak for discussions, their paper appeared later in the conference proceedings and could not compete for the attention of the community.

Later, in 1963, Feynman, who tried to be generous with ideas, publicly stated, "The . . . theory that was discovered by Sudarshan and Marshak, publicized by Feynman and Gell-Mann . . ." But it was too little, too late. It would have been hard in the best of times to compete in the limelight with Feynman and Gell-Mann, and Sudarshan had to live for years with the knowledge that the universal form of the weak interaction, which two of the world's physics heroes had discovered, was first proposed—and with more confidence—by him.

Sudarshan's theory, as elucidated beautifully in Feynman and Gell-Mann's paper, became known as the V-A theory of the weak interaction. The reason for the name is technical and will make more sense in coming chapters, but the fundamental idea is simple, though it sounds both ridiculous and meaningless: the currents in the Fermi theory must be "left-handed."

To understand this terminology, recall that in quantum mechanics elementary particles such as electrons, protons, and neutrinos have spin angular momentum—they behave as if they are spinning even though classically a point particle without extension can't be pictured as spinning. Now, consider the direction of their motion and pretend for a moment the particle is like a top spinning around that axis. Put your right

hand out and let your thumb point in the direction of the particle's motion. Then curl your other fingers around. If they are curling in the same
(counterclockwise) direction that the particle/top is spinning about the
direction of motion, the particle is said to be right-handed. If you put
your left hand out and do the same thing, a left-handed particle would
be spinning clockwise to match the direction of your left-curled hand:

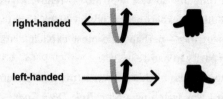

Just as viewing your left hand in a mirror will make it look like a right
hand, if you see a spinning arrow in the mirror, its direction of motion
will be flipped, so that if the arrow is moving away from you in the real
world, it will be moving toward you in the mirror, but the spin will not
be flipped. Thus, in the mirror a left-handed particle will turn into a
right-handed particle. (And so, if the poor souls in Plato's cave had had
mirrors, they might have felt less strange about the shadows of arrows
flipping direction.)

This working picture of left-handed particles is not exact, because if
you think about it, you can also turn a left-handed particle into a right-
handed particle by simply moving faster than the particle. In a frame
in which a person at rest observes the particle zipping by, it may be
moving to the left. But if you hop in a rocket and head off to the left and
pass by that particle, then relative to you, it is moving to the right. As
a result, only for particles that are massless—and are therefore moving
at the speed of light—is the above description exact. For, if a particle
is moving at the speed of light, nothing can move fast enough to pass
the particle. Mathematically, the definition of left-handed has to take
this effect into account, but this complication need not concern us any
more here.

Electrons can spin in either direction, but what the V-A interaction implies mathematically is that only those moving electrons whose currents are left-handed can "feel" the weak force and participate in neutron decay. Right-handed currents don't feel the force.

What is more amazing is that neutrinos only feel the weak force, and no other force. As far as we can tell, neutrinos are *only* left-handed. It is not just that only one sort of neutrino current engages in the weak interaction. In all the experimental observations so far, there *are* no right-handed neutrinos—perhaps the most explicit demonstration of the violation of parity in nature.

The seeming silliness of this nomenclature was underscored to me years ago when I was watching a *Star Trek: Deep Space Nine* episode, during which a science officer on the space station discovers something wrong with the laws of probability in a gaming casino. She sends a neutrino beam through the facility, and the neutrinos are observed to be coming out only left-handed. Clearly something was wrong.

Except that is the way it *really* is.

What is wrong with nature? How come, for at least one of the fundamental forces, left is different from right? And why should neutrinos be so special? The simple answer to these questions is that we don't yet know, even though our very existence, which derives from the nature of the known forces, ultimately depends on it. That is one reason we are trying to find out. The elucidation of a new force led to a new puzzle, and like most puzzles in science, it ultimately provided the key that would lead physicists down a new path of discovery. Learning that nature lacked the left-right symmetry that everyone had assumed was fundamental led physicists to reexamine how symmetries are manifested in the world, and more important, how they are not.

Chapter 13

ENDLESS FORMS MOST BEAUTIFUL: SYMMETRY STRIKES BACK

Now faith is the substance of things hoped for,
the evidence of things not seen.

—HEBREWS 11:1

Borrowing from Pauli, we can say Mother Nature *is* a weak left-hander. With the shocking realization that nature distinguishes left from right, physics itself took a strange left turn down a road with no familiar guideposts. The beautiful order of the periodic table governing phenomena on atomic scales gave way to the mystery of the nucleus and the inscrutable nature of the forces that governed it.

Gone were the seemingly simple days of light, motion, electromagnetism, gravity, and quantum mechanics. The spectacularly successful theory of quantum electrodynamics, which had previously occupied the forefront of physics, seemed to be replaced by a confusing world of exotic phenomena associated with the other two newly discovered weak and strong nuclear forces that governed the heart of matter. Their effects and properties could not easily be isolated, despite that one force was thousands of times stronger than the other. The world of fundamental particles appeared to be ever more complicated, and the situation was getting more confusing with each passing year.

. . .

If the discovery of parity violation created shadows of confusion by demonstrating that nature had completely unexpected preferences, the first rays of light arose from the realization that other nuclear quantities, which on the surface seemed quite different, might, when viewed from a fundamental perspective, be not so different at all.

Perhaps the most important discovery in nuclear physics was that protons and neutrons could convert into each other, as Yukawa had speculated years earlier. This was the basis of the emerging understanding of the weak interaction. But most physicists felt that it was also the key to understanding the strong force that appeared to hold nuclei together.

Two years before his revolutionary work with T.-D. Lee, exposing the demise of the sacred left-right symmetry of nature, C.-N. Yang had concentrated his efforts on trying to understand how a different type of symmetry, borrowed from quantum electrodynamics, might reveal an otherwise hidden beauty inside the nucleus. Perhaps, as Galileo discovered regarding the basis of motion, the most obvious things we observe about nature are also the things that most effectively mask its fundamental properties.

What had slowly become clear, not only from the progress in understanding neutron decay and other weak effects in nuclei, but also from looking at strong nuclear collisions, was that the obvious distinction between protons and neutrons—the proton is charged and the neutron is neutral—might, as far as the underlying physics governing the nucleus is concerned, be irrelevant. Or at least as irrelevant as the apparent distinction between a falling feather and a falling rock is to our understanding of the underlying physics of gravity and falling objects.

First off, the weak force could convert protons into neutrons. More important, when one examined the rates of other, stronger nuclear re-actions involving proton or neutron collisions, replacing neutrons by protons and vice versa didn't significantly change the results.

In 1932, the year the neutron was discovered, Heisenberg had sug-gested that the neutron and proton might be just two states of the same particle, and he invented a parameter he called isotopic spin to distinguish them. After all, their masses are almost the same, and light-stable nuclei contain equal numbers of them. Following this, and after the recognition by the distinguished nuclear physicists Benedict Cassen, Edward Con-don, Gregory Breit, and Eugene Feenberg that nuclear reactions seemed to be largely blind to distinguishing protons and neutrons, the brilliant mathematical physicist Eugene Wigner suggested that isotopic spin was "conserved" in nuclear reactions—implying an underlying symmetry governing the nuclear forces between protons and neutrons. (Wigner had earlier developed rules demonstrating how symmetries in atomic systems ultimately allowed a complete classification of atomic states and the transitions between them, for which he later won the Nobel Prize.)

Earlier, when discussing electromagnetism, I noted that the net electric charge doesn't change during electromagnetic interactions—i.e., electric charge is conserved—because of an underlying symmetry between positive and negative charges. The underlying connection be-tween conservation laws and symmetries is far broader and far deeper than this one example. The deep and unexpected relationship between conservation laws and symmetries of nature has been the single most important guiding principle in physics in the past century.

In spite of its importance, the precise mathematical relationship be-tween conservation laws and symmetries was only made explicit in 1915 by the remarkable German mathematician Emmy Noether. Sadly, al-though she was one of the most important mathematicians in the early twentieth century, Noether worked without an official position or pay for much of her career.

Noether had two strikes against her. First, she was a woman, which made obtaining education and employment during her early career difficult, and second, she was Jewish, which ultimately ended her academic career in Germany and resulted in her exile to the United States shortly before she died. She managed to attend the University of Erlangen as one of 2 female students out of 986, but even then she was only allowed to audit classes after receiving special permission from individual professors. Nevertheless, she passed the graduation exam and later studied at the famed University of Göttingen for a short period before returning to Erlangen to complete her PhD thesis. After working for seven years at Erlangen as an instructor without pay, she was invited in 1915 to return to Göttingen by the famed mathematician David Hilbert. Historians and philosophers among the faculty, however, blocked her appointment. As one member protested, "What will our soldiers think when they return to the university and find that they are required to learn at the feet of a woman?" In a retort that eternally reinforced my admiration for Hilbert, beyond that for his remarkable talent as a mathematician, he replied, "I do not see that the sex of the candidate is an argument against her admission as a *Privatdozent*. After all, we are a university, not a bathhouse."

Hilbert was overruled, however, and while Noether spent the next seventeen years teaching at Göttingen, she was not paid until 1923, and in spite of her remarkable contributions to many areas of mathematics—so many and so deep that she is often considered one of the great mathematicians of the twentieth century—she was never promoted to the position of professor.

Nevertheless, in 1915, shortly after arriving at Göttingen, she proved a theorem that is now known as Noether's theorem, which all graduate students in physics learn, or should learn, if they are to call themselves physicists.

. . .

Returning once again to electromagnetism, the relationship between the arbitrary distinction between positive and negative (had Benjamin Franklin had a better understanding of nature when he defined positive charge, electrons would today probably be labeled as having positive, not negative, charge) and the conservation of electric charge—namely, that the total charge in a system before and after any physical reaction doesn't change— is not at all obvious. It is in fact a consequence of Noether's theorem, which states that for every fundamental symmetry of nature—namely for every transformation under which the laws of nature appear unchanged—some associated physical quantity is conserved. In other words, some physical quantity doesn't change over time as physical systems evolve. Thus:

- The conservation of electric charge reflects that the laws of nature don't change if the sign of all electric charges is changed.
- The conservation of energy reflects that the laws of nature don't change with time.
- The conservation of momentum reflects that the laws of nature don't change from place to place.
- The conservation of angular momentum reflects that the laws of nature don't depend on which direction a system is rotated.

Hence, the claimed conservation of isotopic spin in nuclear reactions is a reflection of the experimentally verified claim that nuclear interactions remain roughly the same if all protons are changed into neutrons and vice versa. It is reflected as well in the world of our experience, in that for light elements, at least, the number of protons and neutrons in the nucleus is roughly the same.

In 1954, Yang, and his collaborator at the time, Robert Mills, went one important step further, once again thinking about light. Electromagnetism and quantum electrodynamics do not just have the simple symmetry that tells us that there is no fundamental difference between negative charge and positive charge, and that the label is arbitrary. As I described at length earlier, a much more subtle symmetry is at work as well, one that ultimately determines the complete form of electrodynamics.

Gauge symmetry in electromagnetism tells us that we can change the definition of positive and negative charge locally without changing the physics, as long as there is a field, in this case the electromagnetic field, that can account for any such local alterations to ensure that the long-range forces between charges are independent of this relabeling. The consequence of this in quantum electrodynamics is the existence of a massless particle, the photon, which is the quantum of the electromagnetic field, and which conveys the force between distant particles.

In this sense, that gauge invariance is a symmetry of nature ensures that electromagnetism has precisely the form it has. The interactions between charged particles and light are prescribed by this symmetry.

Yang and Mills then asked what would happen if one extended the symmetry that implies that we could interchange neutrons and protons everywhere without changing the physics, into a symmetry that allows us to change what we label as "neutron" and "proton" differently from place to place. Clearly by analogy with quantum electrodynamics, some new field would be required to account for and neutralize the effect of these arbitrarily varying labels from place to place. If this field is a quantum field, then could the particles associated with this field somehow play a role in, or even completely determine, the nature of the nuclear forces between protons and neutrons?

These were fascinating questions, and to their credit Yang and Mills didn't merely ask them, they tried to determine the answers by exploring specifically what the mathematical implications of such a new type of gauge symmetry associated with isotopic spin conservation would be.

It became clear immediately that things would get much more complicated. In quantum electrodynamics, merely switching the sign of charges between electrons and positrons does not change the magnitude of the net charge on each particle. However, relabeling the particles in the nucleus replaces a neutral neutron with a positively charged proton. Therefore whatever new field must be introduced in order to cancel out the effect of such a local transformation so that the underlying physics is unchanged must itself be charged. But if the field is itself charged, then, unlike photons—which, being neutral, don't themselves interact directly with other photons—this new field would also have to interact with itself.

Introducing the need for a new charged generalization of the electromagnetic field makes the mathematics governing the theory much more complex. In the first place, to account for all such isotopic spin transformations one would need not just one such field but three fields, one positively charged, one negatively charged, and one neutral. This means that a single field at each point in space, like the electromagnetic field in QED, which points in a certain direction in space with a certain magnitude (and is called a vector field in physics for this reason), is not sufficient. The electric field must be replaced by a field described by a mathematical object called a matrix—not to be confused with anything having to do with Keanu Reeves.

Yang and Mills explored the mathematics behind this new and more complex type of gauge symmetry, which today we call either a nonabelian gauge symmetry—arising from a particular mathematical property of matrices that makes multiplying them different from multiplying numbers—or, in deference to Yang and Mills, a Yang-Mills symmetry.

Yang and Mills's article appears at first glance to be an abstract—or purely speculative—mathematical exploration of the implications of a guess about the possible form of a new interaction, motivated by the observation of gauge symmetry in electromagnetism. Nevertheless, it was not an exercise in pure mathematics. The paper tried to explore possible observable consequences of the hypothesis to see if it might relate to the real world. Unfortunately the mathematics was sufficiently complicated such that the possible observable signatures were not so obvious.

One thing *was* clear, however. If the new "gauge fields" were to account for and thus cancel out the effects of separate isotopic spin transformations made in distant locations, the fields would have to be massless. This is equivalent to saying that only because photons are massless can the force they transmit between particles be arbitrarily long-range. To return to my chessboard analogy, you need a single rule-book to tell you how to properly move over the entire board if I have previously changed the colors of the board randomly from place to place. But having massive gauge fields, which cannot be exchanged over arbitrarily long distances, is equivalent to having a rulebook that tells you how to compensate for changing colors only on nearby squares around your starting point. But this would not allow you to move pieces across the board to distant locations.

In short, a gauge symmetry such as that in electromagnetism, or in the more esoteric Yang-Mills proposal, only works if the new fields required by the symmetry are massless. Amid all the mathematical complexity, this one fact is inviolate.

But we have observed in nature no long-range forces involving the exchange of massless particles other than electromagnetism and gravity. Nuclear interactions are short-range—they only apply over the size of the nucleus.

This obvious problem was not lost on Yang and Mills, who recognized it and, frankly, punted. They proposed that somehow their new particles could become massive when they interacted with the nucleus. When they tried to estimate masses from first principles, they found the theory was too mathematically complicated to allow them to make reasonable estimates. All they knew was that empirically the mass of the new gauge particles would have to be greater than that of pions in order to have avoided detection in then-existing experiments.

Such a willingness to throw their hands in the air might have seemed either lazy or unprofessional, but Yang and Mills knew, as Yukawa had known before them, that no one had been able to write down a sensible quantum field theory of a particle like the photon, but one that, unlike the photon, had a mass. So it didn't seem worthwhile at the time to try to solve all the problems of quantum field theory at once. Instead, with less irreverence than Jonathan Swift, they merely presented their paper as a modest proposal, to spur the imagination of their colleagues.

Wolfgang Pauli, however, would have none of it. While he had thought of some related ideas a year earlier, he had discarded them. Moreover, he felt that all this talk about quantum uncertainties in estimating masses was a red herring. If there *was* to be a new gauge symmetry in nature associated with isotopic spin and governing nuclear forces, the new Yang-Mills particles, like the photon, would have to be massless.

For these reasons, among others, the Yang-Mills paper made far less of a stir at the time than the later Yang and Lee opus. To most physicists it was an interesting curiosity at best, and the discovery of parity violation seemed much more exciting.

But not to Julian Schwinger, who was no ordinary physicist. A child prodigy who had graduated from university by the age of eighteen, he received his PhD by the age of twenty-one. Perhaps no two physicists could have been as different as he and Richard Feynman, who shared the Nobel Prize in 1965 for their separate but equivalent work developing the theory of quantum electrodynamics. Schwinger was refined, formal, and brilliant. Feynman was brilliant, casual, and certainly not refined. Feynman relied often on intuition and guesswork, building on prodigious mathematical skill and experience. Schwinger's mathematical skill was every bit Feynman's equal, but Schwinger worked in an orderly fashion, manipulating complicated mathematical expressions with an ease not possible for ordinary mortals. He joked about Feynman diagrams, which Feynman had developed to make what had previously been perilously laborious calculations in quantum field theory manageable, saying, "Like the silicon chips of more recent years, the Feynman diagram was bringing computation to the masses." Both of them shared one characteristic, however. They marched to the beat of a different drummer . . . in opposite directions.

Schwinger took the Yang-Mills idea seriously. The mathematical beauty must have appealed to him. In 1957, the same year that parity violation was discovered, Schwinger made a bold and seemingly highly unlikely suggestion that the weak interaction responsible for the decay of neutrons into protons, electrons, and neutrinos might benefit from the possibility of Yang-Mills fields, but in a new and remarkable way. He proposed that the observed gauge symmetry of electromagnetism might simply be one part of a larger gauge symmetry in which new gauge particles might mediate the weak interaction that caused neutrons to decay.

An obvious objection to this kind of unification is that the weak interaction is far weaker than electromagnetism. Schwinger had an answer for this. If somehow the new gauge particles were *very* heavy, almost one hundred times heavier than protons and neutrons, then the interaction they might mediate would be of much shorter range than even the size of a nucleus, or even a single proton or neutron. In this case, one could work out that the probability that this interaction would cause a neutron to decay would be small. Thus, if the range of the weak interaction was small, these new fields, the strength of whose intrinsic coupling to electrons and protons on small scales could be comparable to the strength of electromagnetism, could nevertheless, on the scale of nuclei and larger, appear to be much, much weaker.

Put more bluntly, Schwinger proposed the outrageous idea that electromagnetism and the weak interaction were part of a single Yang-Mills theory, in spite of the remarkable and obvious differences between them. He thought that perhaps the photon could be the neutral member of a Yang-Mills-type set of three gauge particles required by treating isotopic spin as a gauge symmetry, with the charged versions conveying the weak interaction and being responsible for mediating the decay of neutrons. Why the charged particles would have a huge mass while the photon was massless, he had no idea. But, as I have often said, lack of understanding is neither evidence for God, nor evidence that one is necessarily wrong. It just is evidence of lack of understanding.

Schwinger was not only a brilliant physicist but a brilliant teacher and mentor. While Feynman had few successful students, probably because none of them could keep up with him, Schwinger seemed to have a knack for guiding brilliant PhD students. In his life he supervised more than seventy PhDs, and four of his students later won the Nobel Prize.

Schwinger was sufficiently interested in relating the weak interaction to electromagnetism that he encouraged one of his dozen graduate students at Harvard at the time to explore the issue. Sheldon Glashow graduated in 1958 with a thesis on the subject and continued to explore the issue for the next few years as a National Science Foundation postdoctoral researcher in Copenhagen. In his Nobel lecture twenty years later, Glashow indicated that he and Schwinger had planned to write a manuscript on the subject after Glashow graduated, but one of them lost the first draft of the manuscript, and they never got back to it.

Glashow was no clone of Schwinger's. Refined and brilliant, yes, but also brash, playful, and boisterous, Glashow did research that was not characterized by mathematical acrobatics, but rather by a keen focus on physical puzzles and exploring new possible symmetries of nature that might resolve them.

When I was a young graduate student in physics at MIT, I was initially drawn to deep mathematical questions in physics and had written my admissions essay for my PhD application on just this subject. Within a few years I found myself depressed by the nature of the mathematical investigations I was pursuing. I met Glashow at a summer school for PhD students in Scotland and became friends with both him and his family—a friendship that continued to blossom when we later became colleagues at Harvard. The year after we met, he spent a sabbatical year at MIT. During this important time for me, when I was considering alternatives, he said to me, "There's physics, and there's formalism, and you have to know the difference." Implicit in this advice was the suggestion that I should pursue physics. When I saw the fun he was having, it became easier to consider joining in.

I soon realized that for me to make progress in physics I needed to work on questions driven primarily by physical issues, not ones driven primarily by mathematical issues. The only way I could do that would be to keep in touch with ongoing experiments—and new experimental results. By watching Shelly and how he did physics, I realized that he had an uncanny ability to know which experiments were interesting, and which results might be significant or might point toward something new. Part of this was undoubtedly innate, but part was based on a lifetime of keeping in touch with what was happening on the ground. Physics is an empirical science, and we lose touch with that at our peril.

In Copenhagen, Glashow realized that if he wanted to properly implement Schwinger's proposal to connect the weak interaction with the electromagnetic interaction, then simply making the photon be the neutral member of a triplet of gauge particles, with the charged members becoming massive by some as yet unknown miracle, wouldn't fly. This couldn't explain the proper nature of the weak interaction, in particular the strange fact that the weak interaction seemed to apply only to left-handed electrons (and neutrinos), whereas electromagnetic interactions don't depend on whether the electrons are left- or right-handed.

The only solution to this problem would be if another neutral gauge particle existed—in addition to the photon—which itself coupled to only left-handed particles. But clearly the new neutral particle would also have to be heavy since the interactions it mediated would have to be weak as well.

Glashow's ideas were reported to the physics community by Murray Gell-Mann at the 1960 Rochester meeting, as Gell-Mann had by then recruited Glashow to Caltech to work in Gell-Mann's group. Glashow's paper on the subject, submitted in 1960, appeared in 1961 in print. Yet, no sudden stampede occurred in response.

After all, two fundamental problems remained with Glashow's proposal. The first was the long-familiar problem of how one could have the different masses of the particles needed to convey the different forces, when gauge symmetries required all the gauge particles to be massless. Glashow simply stated in the introduction of his paper, following in a long line of such hubris, "It is a stumbling block we must overlook."

The second problem was more subtle, but from an experimental perspective equally severe. Neutron decay, pion decay, and muon decay, if they were indeed mediated by some new particles conveying the weak force, all appeared to require only the exchange of new charged particles. No weak interaction had been observed that would require the exchange of a new neutral particle. If such a new neutral particle did exist, calculations at the time suggested it would allow the other known heavier mesons that decayed into two or three pions (and were responsible for the original confusion that led to the discovery of parity violation) to decay much more rapidly than they were observed to decay.

For these reasons, Glashow's proposal drifted into the background as physicists became entranced with the new particle zoo that was emerging out of accelerators, and the concomitant opportunity for new discoveries. Yet several of the key theoretical ingredients needed to complete a revolution in fundamental physics were in place, but it was far from obvious at the time. That within slightly more than a decade after Glashow's paper was published all of the known forces in nature save gravity would be unveiled and understood would have seemed like pure fantasy at the time.

And symmetry would be the key.

Chapter 14

COLD, STARK REALITY:
BREAKING BAD OR BEAUTIFUL?

From whose womb has come the ice? And the frost
of heaven, who has given it birth?

—JOB 38:29

It is easy to pity the poor protagonists in Plato's cave, who may understand everything there is to know about the shadows on the wall, except that they are shadows. But appearances can be deceiving. What if the world around us is just a similar shadow of reality?

Imagine, for example, that you wake up one cold winter morning and look out your window, and the view is completely obscured by beautiful ice crystals, forming strange patterns on the glass. It might look like this:

Photograph by Helen Filatova

The beauty of the image is striking at least in part because of the re-markable order on small scales lurking within the obvious randomness on large scales. Ice crystals have grown gorgeous treelike patterns, start-ing in random directions and bumping into each other at odd angles. The dichotomy between small-scale order and large-scale randomness suggests that the universe would look very different to tiny physicists or mathematicians confined to live on the spine of one of the ice crystals in the image.

One direction in space, corresponding to the direction along the spine of the ice crystal, would be special. The natural world would ap-pear to be oriented around that axis. Moreover, given the crystal lattice structure, electric forces along the spine would appear to be quite dif-ferent from the forces perpendicular to it: the forces would behave as if they were different forces.

If the physicist or mathematician living on the crystal was clever, or, like the mathematician in Plato's cave, lucky enough to leave the crystal, it would soon become clear that the special direction that governed the physics of the world they were used to was an illusion. They would find, or surmise, that other crystals could point in many other directions. Ultimately if they could observe the window from the outside on large enough scales, the underlying symmetry of nature under rotations in all directions, reflected in the growth of the crystals in all directions, would become manifest.

The notion that the world of our experience is a similar accident of our particular circumstances rather than a direct reflection of underly-ing realities has become central to modern physics. We even give it a fancy name: spontaneous symmetry breaking.

I mentioned one sort of spontaneous symmetry breaking earlier when discussing parity, or left-right symmetry. Our left hands look dif-ferent from our right hands even though electromagnetism—the force that governs the building of large biological structures such as our bod-ies—doesn't distinguish between left and right.

Two other examples I know of, both presented by distinguished physicists, also help illuminate spontaneous symmetry breaking in different ways that might be useful. Abdus Salam, who won a Nobel Prize in 1979 for work that depended crucially on this phenomenon, described a situation that is familiar to all of us: sitting down with a group of people at a round dining table set for, say, eight people. When you sit down, it may not be obvious which wineglass is yours and which is your neighbor's—the one on the right or the one on the left. But regardless of the laws of etiquette, which dictate it should be on your right, once the first person picks up her glass, everyone else at the table has only one option if everyone is to get a drink. Even though the underlying symmetry of the table is manifest, the symmetry gets broken when a direction is chosen for the wineglasses.

Yoichiro Nambu, another Nobelist who was the first physicist to describe spontaneous symmetry breaking in particle physics, gave another example that I will adapt here. Take a rod, or even a drinking straw, hold it up with one end on a table, and press down on the top end of the rod. Ultimately the rod will bend. It could bend in any direction, and if you try the experiment several times, you may find it bending in different directions each time. Before you press down, the rod has complete cylindrical symmetry. Afterward, one direction among many possibilities has been chosen, not determined by the underlying physics of the rod but by the accident of the particular way you press on the rod each time. The symmetry has been broken spontaneously.

If we now return to the world of the frozen window, the characteristics of materials can change as we cool systems down. Water freezes, gases liquefy, and so on. In physics, such a change is called a phase transition, and as the window example demonstrates, whenever a system undergoes a phase transition, it is not unusual to find that symmetries associated with one phase will disappear in the other phase. Before the ice froze into the crystals on the window, the water droplets wouldn't have been so ordered, for example.

One of the most astonishing phase transitions ever witnessed in science was first observed by the Dutch physicist Kamerlingh Onnes on April 8, 1911. Onnes had—remarkably—been able to cool materials to temperatures never before achieved, and he was the first person to liquefy helium, at just four degrees above absolute zero. For this experimental prowess he was later awarded a Nobel Prize. On April 8, when cooling a mercury wire down to 4.2 degrees above absolute zero in a liquid helium bath and measuring its electrical resistance, to his astonishment he discovered that the resistance suddenly dropped to zero. Currents could flow in the wire indefinitely once they began, even after any battery that started the flow was removed. Demonstrating that his talent for public relations was as astute as his experimental talents, he coined the term *superconductivity* to describe this remarkable and completely unexpected result.

Superconductivity was so unexpected and strange that it would take almost fifty years after the discovery of quantum mechanics, on which it depends, before a fascinating physics explanation was developed by the team of John Bardeen, Leon Cooper, and Robert Schrieffer, in 1957. (That was same year that parity violation was observed, and that Schwinger proposed a model to try to unify the weak and electromagnetic interaction.) Their work was a tour de force, built on a succession of insights made over several decades of work. Ultimately the explanation relies on an unexpected phenomenon that can only occur in certain materials.

In empty space, electrons repel other electrons because like charges repel each other. However, in certain materials, as they are cooled, electrons can actually bind to other electrons. This happens in the material because a free electron tends to attract around it positively charged ions. If the temperature is extremely low, then another electron can be attracted to the positively charged field around the first electron. Pairs of electrons can bind together, with the glue, if you wish, being the positively charged field caused by the attraction of the first electron on the lattice of positive charges associated with the atoms in the material.

Since the nuclei of atoms are heavy and pinned in place by relatively strong atomic forces, the first electron slightly distorts the lattice of nearby atoms, moving some of the atoms slightly closer to the electron than they would otherwise be. Distortions of the lattice in general cause vibrations, or sound waves, in the material. In the quantum world these vibrations are quantized and are called phonons. Leon Cooper discovered that these phonons can bind pairs of electrons, as I have described above, so these are called Cooper pairs.

The true magic of quantum mechanics occurs next. When mercury (or any of several other materials) is cooled below a certain point, a phase transition occurs and all the Cooper pairs suddenly coalesce into a single quantum state. This phenomenon, called Bose-Einstein condensation, occurs because unlike fermions, particles with integral quantum mechanical spin, such as photons, or even particles with zero spin, instead prefer to all be in the same state. This was proposed first by the Indian physicist Satyendra Nath Bose and later elaborated upon by Einstein. Once again light played a crucial role, as Bose's analysis involved the statistics of photons, and Bose-Einstein condensation is intimately related to the physics governing lasers, in which many individual photons all behave coherently in the same state. For this reason particles with integral spin such as photons are called bosons, to distinguish them from fermions.

In a gas or a solid at room temperature, normally so many collisions occur between particles that their individual states are changing rapidly and any collective behavior is impossible. However, a gas of bosons can coalesce at a low enough temperature into a Bose-Einstein condensate, in which the individual particles' identities disappear. The whole system behaves like a single, sometimes macroscopic, object, but in this case acting via the rules of quantum mechanics, rather than classical mechanics.

As a result, a Bose-Einstein condensate can have exotic properties, the way laser light can behave very differently from normal light coming from flashlights. Since a Bose-Einstein condensate is a huge amalgamation of what would otherwise be individual noninteracting particles, now tied together into a single quantum state, creating such a condensate required exotic and special atomic physics experiments. The first direct observation of such a condensation from a gas of particles did not take place until 1995, by the US physicists Carl Wieman and Eric Cornell, another feat that was deemed worthy of a Nobel Prize.

What makes the possibility of such a condensation inside bulk materials such as mercury so strange is that the fundamental particles initially involved are electrons—which not only normally repel other electrons, but in addition have spin ½ and, as fermions, have precisely the opposite behavior of bosons, as I described above.

But when the Cooper pairs form, the two electrons each act in concert, and since both of them have spin ½, the combined object has integral (2 × ½) spin. Voilà, a new kind of boson is created. The lowest-energy state of the system, to which it relaxes at low temperature, is a condensate of Cooper pairs—all condensed into a single state. When that happens, the properties of the material change completely.

Before the condensate forms, when a voltage is applied to a wire, individual electrons begin to move to form an electric current. As they bump into atoms along the way, they dissipate energy, producing an electrical resistance that we are all familiar with, and heating up the wire. Once the condensate forms, however, the individual electrons and even each Cooper pair no longer have any individual identity. Like the Borg in *Star Trek*, they have assimilated into a collective. When a current is applied, the whole condensate moves as one entity.

Now, if the condensate were to bounce off an individual atom, the trajectory of the whole condensate would change. But this would take a lot of energy, much more than would have been required to redirect the flow of an individual electron. Classically we can think of the result as follows: at low temperatures, not enough heat energy is available in the random jittering of atoms to cause a change of motion of the bulk condensate of particles. It would again be like trying to move a truck by throwing popcorn at it. Quantum mechanically the result is similar. In this case we would say that to change the configuration of the condensate would require the whole condensate of particles to shift by a large fixed amount to a new quantum state that differs in energy from the state it is in. But no such energy is available from the thermal bath at low temperature. Alternatively, we might wonder if the collision could break apart two electrons from a Cooper pair in the condensate—sort of like knocking off the rearview mirror when a truck collides with a post. But at low temperatures everything is moving too slowly for that to happen. So the current flows unimpeded. The Borg would say, resistance is futile. But in this case resistance is simply nonexistent. A current, once initiated, will flow forever, even if the battery initially attached to the wire is removed.

This was the Bardeen-Cooper-Schrieffer (BCS) theory of superconductivity, a remarkable piece of work, which ultimately explained all of the experimental properties of superconductors such as mercury. These new properties signal that the ground state of the system has changed from what it had been before it became a superconductor, and like ice crystals on a window, these new properties reflect spontaneous symmetry breaking. In superconductors the breaking of symmetry is not as visually obvious as it is in the ice crystals on a windowpane, but it is there, under the surface.

Mathematically, the signature of this symmetry breaking is that suddenly, once the condensate of Cooper pairs forms, a large minimum energy is now required to change the configuration of the whole material. The condensate acts like a macroscopic object with some large mass. The generation of such a "mass gap" (as it is called—expressed as the minimum energy it takes to break the system out of its superconducting state) is a hallmark of the symmetry-breaking transition that produces a superconductor.

You might be wondering what all of this, as interesting as it might be, has to do with the story we have been focusing on, namely understanding the fundamental forces of nature. With the benefit of hindsight, the connection will be clear. However, in the tangled and confused world of particle physics in the 1950s and '60s the road to enlightenment was not so direct.

In 1956, Yoichiro Nambu, who had recently moved to the University of Chicago, heard a seminar by Robert Schrieffer on what would become the BCS theory of superconductivity, and it left a deep impression on him. He, like most others interested in particle physics at the time, had been wrestling with how the familiar particles that make up atomic nuclei—protons and neutrons—fit within the particle zoo and the jungle of interactions associated with their production and decay.

Nambu, like others, was struck by the almost identical masses of the proton and the neutron. It seemed to him, as it had to Yang and Mills, that some underlying principle in nature must produce such a result. Nambu, however, speculated that the example of superconductivity might provide a vital clue—in particular the appearance of a new characteristic energy scale associated with the excitation energy required to break apart the Cooper-pair condensate.

For three years Nambu explored how to adapt this idea to symmetry breaking in particle physics. He proposed a model by which a similar condensate of some fields that might exist in nature and the minimum energy to create excitations out of this condensate state could be characteristic of the large mass/energy associated with protons and neutrons.

Independently, he and the physicist Jeffrey Goldstone discovered that a hallmark of such symmetry breaking would be the existence of other massless particles, now called Nambu-Goldstone (NG) bosons, whose interactions with other matter would also reflect the nature of the symmetry breaking. An analogy of sorts can be made here to a more familiar system such as an ice crystal. Such a system spontaneously breaks the symmetry under spatial translation because moving in one direction things look very different from when moving in another direction. But in such a crystal, tiny vibrations of individual atoms in the crystal about their resting positions are possible. These vibrational modes—called phonons, as I have mentioned—can store arbitrarily small amounts of energy. In the quantum world of particle physics, these modes would be reflected as Nambu-Goldstone massless particles, because where the equivalence between energy and mass is manifest, excitations that carry little or no energy correspond to massless particles.

And, lo and behold, the pions discovered by Powell closely fit the bill. They are not exactly massless, but they are much lighter than all other strongly interacting particles. Their interactions with other particles have the characteristics one would expect of NG bosons, which might exist if some symmetry-breaking phenomenon existed in nature with a scale of excitation energy that might correspond to the mass/energy scale of protons and neutrons.

But, in spite of the importance of Nambu's work, he and almost all of his colleagues in the field overlooked a related but much deeper consequence of the spontaneous symmetry breaking in the theory of superconductivity that later provided the key to unlock the true mystery of the strong and weak nuclear forces. Nambu's focus on symmetry breaking was inspired, but the analogies that he and others drew to superconductivity were incomplete.

It seems that we are much closer to the physicists on that ice crystal on the windowpane than we ever imagined. But just as one might imagine would be the case for those physicists, this myopia was not immediately obvious to the physics community.

Chapter 15

LIVING INSIDE A SUPERCONDUCTOR

*Everyone lies to their neighbor; they flatter with their
lips but harbor deception in their hearts.*

—PSALMS 12:2

The mistakes of the past may seem obvious with the benefit of hindsight, but remember that objects viewed in the rearview mirror are often closer than they appear. It is easy to castigate our predecessors for what they missed, but what is confusing to us today may be obvious to our descendants. When working on the edge, we travel a path often shrouded in fog.

The analogy to superconductivity first exploited by Nambu is useful, but largely for reasons very different from what Nambu and others imagined at the time. In hindsight the answer may seem almost obvious, just as the little clues that reveal the murderer in Agatha Christie stories are clear after the solution. But, as in her mysteries, we also find lots of red herrings, and these blind alleys make the eventual resolution even more surprising.

We can empathize with the confusion in particle physics at the time. New accelerators were coming online, and every time a new collision-energy threshold was reached, new strongly interacting cousins of neutrons and protons were produced. The process seemed as if it would be endless. This embarrassment of riches meant that both theorists and experimentalists were driven to focus on the mystery of the strong nuclear force, which seemed to be where the biggest challenge to existing theory lay.

A potentially infinite number of elementary particles with ever-higher masses seemed to characterize the microscopic world. But this was incompatible with all the ideas of quantum field theory—the successful framework that had so beautifully provided an understanding of the relativistic quantum behavior of electrons and photons.

Berkeley physicist Geoffrey Chew led the development of a popular, influential program to address this problem. Chew gave up the idea that any truly fundamental particles exist and also gave up on any microscopic quantum theory that involved pointlike particles and the quantum fields associated with them. Instead, he assumed that all of the observed strongly interacting particles were not pointlike, but complicated, bound states of other particles. In this sense, there could be no reduction to primary fundamental objects. In this Zen-like picture, appropriate to Berkeley in the 1960s, all particles were thought to be made up of other particles—the so-called bootstrap model, in which no elementary particles were primary or special. So this approach was also called nuclear democracy.

While this approach captivated many physicists who had given up on quantum field theory as a tool to describe any interactions other than the simple ones between electrons and photons, a few scientists were sufficiently impressed by the success of quantum electrodynamics to try to mimic it in a theory of the strong nuclear force—or strong interaction, as it has become known—along the lines earlier advocated by Yang and Mills.

One of these physicists, J. J. Sakurai, published a paper in 1960 rather ambitiously titled "Theory of Strong Interactions." Sakurai took the

Yang-Mills suggestion seriously and tried to explore precisely which photonlike particles might convey a strong force between protons and neutrons and the other newly observed particles. Because the strong interaction was short-range—spanning just the size of the nucleus at best—it seemed the particles required to convey the force would be massive, which was incompatible with any exact gauge symmetry. But otherwise, they would have many properties similar to the photon's, having spin 1, or a so-called vector spin. The new predicted particles were thus dubbed massive vector mesons. They would couple to various currents of strongly interacting particles similar to the way photons couple to currents of electrically charged particles.

Particles with the general properties of the vector mesons predicted by Sakurai were discovered experimentally over the next two years, and the idea that they might somehow yield the secret of the strong interaction was exploited to try to make sense of the otherwise complex interactions between nucleons and other particles.

In response to this notion that some kind of Yang-Mills symmetry might be behind the strong interaction, Murray Gell-Mann developed an elegant symmetry scheme he labeled in a Zen-like fashion the Eightfold Way. It not only allowed a classification of eight different vector mesons, but also predicted the existence of thus-far-unobserved strongly interacting particles. The idea that these newly proposed symmetries of nature might help bring order to what otherwise seemed a hopeless menagerie of elementary particles was so exciting that, when his predicted particle was subsequently discovered, it led to a Nobel Prize for Gell-Mann.

But Gell-Mann is remembered most often for a more fundamental idea. He, and independently George Zweig, introduced what Gell-Mann called *quarks*—a word borrowed from James Joyce's *Finnegans Wake*—which would physically help explain the symmetry properties of his Eightfold Way. If quarks, which Gell-Mann viewed simply as a nice mathematical accounting tool (just as Faraday had earlier viewed his proposal of electric and magnetic fields), were imagined to comprise all strongly interacting particles such as protons and neutrons, the symmetry and properties of the known particles could be predicted. Once again, the smell of a grand synthesis that would unify diverse particles and forces into a coherent whole appeared to be in the air.

I cannot stress how significant the quark hypothesis was. While Gell-Mann did not advocate that his quarks were real physical particles inside protons and neutrons, his categorization scheme meant that symmetry considerations might ultimately determine the nature not only of the strong interaction, but of all fundamental particles in nature.

However, while one sort of symmetry might govern the structure of matter, the possibility that this symmetry might be extended to some kind of Yang-Mills gauge symmetry that would govern the forces between particles seemed no closer. The nagging problem of the observed masses of the vector mesons meant that they could not truly reflect any underlying gauge symmetry of the strong interaction in a way that could unambiguously determine its form and potentially ensure that it made quantum-mechanical sense. Any Yang-Mills extension of quantum electrodynamics required the new photonlike particles to be massless. Period.

Faced with this apparent impasse, an unexpected wake-up call from superconductivity provided another, more subtle, and ultimately more profound, possibility.

The first person to stir the embers was a theorist who worked directly in the field of condensed matter physics associated with superconductivity in materials. Philip Anderson, at Princeton, later a Nobel laureate for other work, suggested that one of the most fundamental, ubiquitous phenomena in superconductors might be worth exploring in the context of particle physics.

One of the most dramatic demonstrations one can perform with superconductors, especially the new high-temperature superconductors that allow superconductivity to become manifest at liquid-nitrogen temperatures, is to levitate a magnet above the superconductor as shown below:

Creative Commons/Photograph by Mai-Linh Doan

This is possible for a reason discovered in an experiment in 1933 by Walther Meissner and colleagues, explained by theorists Fritz and Heinz London two years later, which goes by the name the Meissner effect.

As Faraday and Maxwell discovered sixty years earlier, electric charges respond in different ways to magnetic and electric fields. In particular, Faraday discovered that a changing magnetic field can cause a current to flow in a distant wire. Equally important, but which I didn't emphasize earlier, is that the resulting current will flow in a way that produces a new magnetic field in a direction that counters the changing external magnetic field. Thus, if the external field is decreasing, the current generated will produce a magnetic field that counters that decrease. If it is increasing, the current generated will be in an opposite direction, producing a magnetic field that works to counter that increase.

You may have noticed that when you are talking on your cell phone and get in certain elevators, particularly ones in which the outer part of the elevator cage is encased in metal, when the door closes your call gets dropped. This is an example of something called a Faraday cage. Since the phone signal is being received as an electromagnetic wave, the metal shields you from the outside signal because currents flow in the metal in a way that counters the changing electric and magnetic fields in the signal, diminishing its strength inside the elevator.

If you had a perfect conductor, with no resistance, the charges in the metal could essentially cancel any effects of the outside changing electromagnetic field. No signal of these changing fields—i.e., no telephone signal—would remain to be detected inside the elevator. Moreover, a perfect conductor will also shield out the effects of any constant external electric field, since the charges can realign in the superconductor in response to any field and completely cancel it out.

But the Meissner effect goes beyond this. In a superconductor, all magnetic fields—even constant magnetic fields such as those due to the magnet above—cannot penetrate into the superconductor. This is because, when you slowly bring a magnet in closer from a large distance, the superconductor generates a current to counter the changing magnetic field that increases as the magnet approaches. But since the material is superconducting, the current continues to flow and does not stop if you stop moving the magnet. Then as you bring the magnet in closer, a larger current flows to counter the new increase. And so on. Thus, because electric currents can flow without dissipation in a superconductor, not only are electric fields shielded, but so are magnetic fields. This is why magnets levitate above superconductors. The currents in the superconductor expel the magnetic field due to the external magnet, and this repels the magnet just as if another magnet were at the surface of the superconductor with north pole facing north pole or south pole facing south pole.

The London brothers, who first attempted to explain the Meissner effect, derived an equation describing this phenomenon inside a superconductor. The result was suggestive. Each different type of superconductor would create a unique characteristic length scale below the surface of the superconductor—determined by the microscopic nature of the supercurrents that are created to compensate any external field—and any external magnetic field would be canceled on this length scale. This is called the London penetration depth. The depth is different for different superconductors and depends on their detailed microphysics in a way the brothers couldn't determine since they didn't have a microscopic theory of superconductivity at the time.

Nevertheless, the presence of a penetration depth is striking because it implies that the electromagnetic field behaves differently inside a superconductor—it is no longer long-range. But if electromagnetic fields become short-range inside the surface, then the carrier of electromagnetic forces must behave differently. The net effect? The photon behaves as if it has mass inside the superconductor.

In superconductors, virtual photons—and the electric and magnetic fields they mediate—can only propagate below the surface through a distance comparable to the London penetration depth, just as would be the case if electromagnetism inside the superconductor resulted from the exchange of massive—not massless—photons.

Now imagine what it would be like to live inside a superconductor. To you, electromagnetism would be a short-range force, photons would be massive, and all the familiar physics that we associate with electromagnetism as a long-range force would disappear.

I want to emphasize how remarkable this is. No experiment you could perform within the superconductor, as long as it remained superconducting, would reveal that photons are massless in the outside world. If you were Plato's philosopher inside such a superconductor, you would have to intuit an incredible amount about the outside world before you could infer that a mysterious and invisible phenomenon was the cause of an illusion. It might take several thousand years of thinking and experiment before you or your descendants could guess the nature of the reality underlying the shadow world in which you live, or before you could build a device with enough energy to break apart Cooper pairs and melt the superconducting state, restoring electromagnetism to its normal form, and revealing the photon to be massless.

In retrospect, we physicists might have expected, just on the grounds of symmetry, and without considering the Meissner effect directly, that photons should behave as massive particles inside a superconductor. The Cooper-pair condensate, being made of electron pairs, has a net electric charge. This breaks the gauge symmetry of electromagnetism because in this background any positive charges one adds to the material will behave differently from negative charges added to the material. So now there is a real distinction between positive and negative. But recall that the masslessness of photons is a sign that the electromagnetic field is long-range, and the long-range nature of the electromagnetic field reflects that it allows local variations in the definition of electric charge in one place to *not* affect the physics globally throughout the material. But if gauge invariance is gone, then local variations in the definition of electric charge will have a real physical effect, so there can be no such long-range field that cancels out such variations. One way to get rid of a long-range field is to make the photon massive.

Now the $64,000 question: Could something like this happen in the world in which we find ourselves living? Could the masses of heavy photonlike particles arise because we are actually living in something akin to a cosmic superconductor? This was the fascinating question that Anderson raised, at least by analogy with regular superconductors.

Before we can answer this question, we need to understand a technical bit of wizardry that allows the generation of mass for a photon in a superconductor.

Recall that in an electromagnetic wave the electric (E) and magnetic (B) fields oscillate back and forth in directions that are perpendicular to the direction of the wave, as shown:

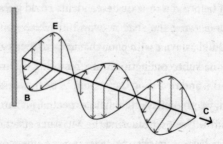

Since there are two perpendicular directions, one could draw an electromagnetic wave in two ways. The wave could look like that shown above, or one could interchange the E and B fields. This reflects that electromagnetic waves have two degrees of freedom, which are called two different polarizations.

This arises from the gauge invariance of electromagnetism, or equivalently from the masslessness of photons. If, however, photons had a mass, then not only would gauge invariance be broken, but a third possibility can arise. The electric and magnetic fields could oscillate *along* the direction of motion, instead of just oscillating perpendicular to this direction. (Since the photons will no longer be traveling at the speed of light, oscillations along the direction of motion of the particles become possible.)

But this means that the corresponding massive photons would have three degrees of freedom, not just two. How can photons pick up this extra degree of freedom in superconductors?

Anderson explored this issue in superconductors, and its resolution is intimately related to a fact that I described earlier. In the absence of electromagnetic interactions in a superconductor, it's possible to produce slight spatial variations in the Cooper-pair condensate that would have arbitrarily small energy cost because Cooper pairs would not interact with each other. However, when electromagnetism is taken into account, those low-energy modes (which would destroy superconductivity) disappear precisely because of the interactions of the charges in the condensate with the electromagnetic field. That interaction causes photons in the superconductor to behave as if they are massive. The new polarization mode of the massive photons in the superconductor comes about as the condensate oscillates in response to the passing electromagnetic wave.

In particle physics language, the massless Nambu-Goldstone modes that correspond to the particle version of the otherwise vanishingly small energy oscillations in the condensate get "eaten" by the electromagnetic field, giving photons a mass, and a new degree of freedom, making the electromagnetic force short-range in the superconductor.

Anderson suggested that this phenomenon—whereby the otherwise massless photon disappears in superconductors and the otherwise massless Nambu-Goldstone mode also disappears, and the two combine to produce a massive photon—might be relevant for the long-standing problem of creating massive Yang-Mills photonlike particles that might be associated with strong nuclear forces.

Anderson stopped short at this point and left hanging the suggestion that this mechanism, motivated by analogy to superconductors, might be applicable in particle theory. Just as when Nambu had stopped short by considering spontaneous symmetry breaking in particle physics using the analogy of superconductivity but did not exploit the phenomenon associated with superconductivity that Anderson later focused on—the Meissner effect that gives mass to photons in superconductors—the explicit application of all these ideas to particle physics was yet to occur.

As a result, the possible profound implications of superconductivity for understanding fundamental particle physics were not immediately recognized by the physics community and remained hidden in the shadows.

Still, the notion that we might live in some kind of cosmic superconductor stretches credulity. After all, humans are capable of generating wild stories to explain what is otherwise not understood, inventing fantastical and hidden causes, such as gods and demons. Was the claimed existence of some hidden condensate of fields throughout space to explain the nature of what were otherwise inexplicable strong nuclear forces any more plausible?

Chapter 16

THE BEARABLE HEAVINESS OF BEING: SYMMETRY BROKEN, PHYSICS FIXED

Gather up the fragments that remain, that
nothing be lost.

—JOHN 6:12

There is remarkable poetry in nature, as there often is in human dramas. And in my favorite epic poems from ancient Greece, written even as Plato was writing about his cave, there emerges a common theme: the discovery of a beautiful treasure previously hidden from view, unearthed by a small and fortunate band of unlikely travelers, who, after its discovery, are changed forever.

Oh, to be so lucky. That possibility drove me to study physics, because the romance of possibly discovering some new and beautiful hidden corner of nature for the first time had an irresistible allure. This story is all about those moments when the poetry of nature merges with the poetry of human existence.

Much poetry exists in almost every aspect of the episodes I am about to describe, but to see it clearly requires the proper perspective. Today, in the second decade of the twenty-first century, we might easily agree about which of the great theories of the twentieth century are most beautiful. But to appreciate the real drama of the progress of science, one has to understand that, at the time they are proposed, beautiful theories often aren't as seductive as they are years later—like a fine wine, or a distant love.

So it was that the ideas of Yang and Mills, and Schwinger and the rest, based on the mathematical poetry of gauge symmetry, failed at the time to inspire or compete with the idea that quantum field theory, with quantum electrodynamics as its most beautiful poster child, wasn't a productive approach to describe the other forces in nature—the weak and strong nuclear forces. For forces such as these, operating on short ranges appropriate to the scale of atomic nuclei, many felt that new rules must apply, and that the old techniques were misplaced.

So too the subsequent attempts by Nambu and Anderson to apply ideas from the physics of materials—called many-body physics, or condensed matter physics—to the subatomic realm were dismissed by many particle physicists, who deeply distrusted whether this emerging field could provide any new insights for "fundamental" physics. The skepticism in the community was expressed by the delightful theorist Victor Weisskopf, who was reported to have said at a seminar at Cornell, "Particle physicists are so desperate these days that they have to borrow from the new things coming up in many-body physics. . . . Perhaps something will come of it."

There was some basis for the skepticism. Nambu had, after all, argued that spontaneous symmetry breaking might explain the large and similar masses of protons and neutrons, and he hoped it might do so while explaining why the pion was so much lighter. But the ideas he borrowed had at their foundation the understanding that the hallmark of spontaneous symmetry breaking was the existence of exactly massless, not *very light*, particles.

Anderson's work was also interesting, to be sure. But because it was written down in the context of a nonrelativistic condensed matter setting—combined with its violating Goldstone's theorem from particle physics, which implied that symmetry breaking and massless particles were inseparable—meant that his claim that massless states disappeared in his example—in electromagnetism in superconductors—was largely also ignored by particle physicists.

Julian Schwinger, however, had not given up the idea that a Yang-Mills gauge theory might explain nuclear forces, and he had continued to argue that the Yang-Mills versions of photons could be massive, albeit without demonstrating how this could come to pass.

Schwinger's work caught the attention of a mild-mannered young British theorist, Peter Higgs, who was then a lecturer in mathematical physics at the University of Edinburgh. A gentle soul, no one would imagine him to be a revolutionary. But reluctant revolutionary he was, although, due to some shortsighted journal editors, he almost didn't get the chance.

In 1960 Higgs had just taken up his post and had been asked to serve on the committee that coordinated the first Scottish Universities Summer School in Physics. This became a venerable school, devoted to different areas of physics. Every four years or so, during three weeks, advanced graduate students and young postdocs would attend lectures on particle physics by senior scientists amid meals lubricated by fine wine and, afterward, hearty whiskey. Among the students that year were the future Nobelists Sheldon Glashow and Martinus Veltman, and Nicola Cabibbo, who in my opinion should also have won the prize. Apparently Higgs, who had been made the wine steward, noticed that these three students never made the morning lectures. They apparently spent the evenings debating physics while drinking wine that they sneaked out of the dining room during meals. Higgs didn't have the opportunity to join the discussions then and therefore didn't learn from Glashow about his novel proposal for unifying the electromagnetic and weak forces, which he had already submitted for publication.

The Scottish summer schools have a poetry of their own. They rotate around the country and periodically return to the beautiful coastal city of St. Andrews, right next to the famous Old Course, the birthplace of golf. In 1980 at St. Andrews, Glashow, fresh from having won a Nobel Prize, and Gerardus 't Hooft, a famous former student of Veltman's, lectured at the school, and I was privileged to attend as a graduate student.

I arrived late and got the smallest room, up in an attic overlooking the Old Course, and enjoyed not only the physics, but also the alcohol, as well as being fleeced for free drinks by one of the lecturers, Oxford physicist Graham Ross, at a miniature-golf putting range next door nicknamed the Himalayas, for good reason. Besides being a physicist of almost otherworldly ability, 't Hooft is also a remarkable artist. He won the 1980 summer school's annual T-shirt design contest, and I still have my autographed 't Hooft T-shirt. Can't bear to part with it, even as eBay beckons. (Twenty years after that program, in 2000, I returned to the summer school, but this time as a lecturer. Unlike Glashow, 't Hooft, Veltman, and Higgs, I didn't return with a Nobel Prize, but I finally got to wear a kilt. Another bucket-list item ticked.)

Following Higgs's stint at the summer school in 1960, he began to study the literature on symmetry and symmetry breaking, examining the work of Nambu, Goldstone, Salam, Weinberg, and Anderson. Higgs became depressed by the seemingly hopeless task of reconciling Goldstone's theorem with the possibility of massive Yang-Mills vector particles that might mediate the strong force. Then in 1964, the magical year when Gell-Mann introduced quarks, Higgs read two papers that gave him hope.

First was a paper by Abraham Klein and Ben Lee—who, before he died in a car crash while driving to a physics meeting, was one of the brightest upcoming particle physicists in the world. They suggested a way to avoid Goldstone's theorem and get rid of otherwise unobserved massless particles in quantum field theories.

Next, Walter Gilbert, a young physicist at Harvard who would soon decide to leave the confusion dominating particle physics for the greener pastures of molecular biology—where he too would win a Nobel Prize, in this case for helping to develop DNA-sequencing techniques—wrote a paper showing that the proposed solution of Klein and Lee's appeared to introduce a conflict with relativity and therefore was suspect.

As we've seen, gauge theories have the interesting property that you can arbitrarily change the definition of positive versus negative charges at each point in space without changing any of the observable physical properties of the system, as long as you allow the electromagnetic field to have the interactions it has and to also change in a way that properly accounts for this new local variation. As a result, you can perform mathematical calculations in any gauge—that is, using any specific local definitions of charges and fields consistent with the symmetry. A symmetry transformation will take you from one gauge to another.

Even though the theory might look quite different in these different gauges, the symmetry of the theory ensures that calculations of any physically measurable quantity are independent of the gauge choice—namely that the apparent differences are illusions that do not reflect the underlying physics that determines the measured values of all physically observable quantities. Thus one could choose whichever gauge made the calculation easier to do and expect to arrive at the same predictions for physically observable quantities by calculating in any other gauge.

As Higgs read Schwinger's papers, Higgs realized that some gauge choices could appear to have the same conflict with relativity that Gilbert had pointed out as plaguing Klein and Lee's proposal. But this apparent conflict was simply an artifact of that choice of gauge. In other gauges it disappeared. Therefore it didn't reflect any real conflict with relativity when it came to making physical predictions that could be tested. Maybe in a gauge theory Klein and Lee's proposal for getting rid of massless particles associated with spontaneous symmetry breaking might be workable after all.

Higgs concluded that spontaneous symmetry breaking in a quantum field theory setting involving a gauge symmetry might obviate Goldstone's theorem and produce a mass for vector bosons that might mediate the strong nuclear force without any leftover massless particles. This would correlate with Anderson's finding of electromagnetism in superconductors in the nonrelativistic case. In other words, the strong force could be a short-range force because of spontaneous symmetry breaking.

Higgs worked for a weekend or two to write down a model adding electromagnetism to the model Goldstone had used to explore spontaneous symmetry breaking. Higgs found just what he had expected: the otherwise massless mode that would have been predicted by Goldstone's theorem became instead the additional polarization degree of freedom of a now massive photon. In other words, Anderson's nonrelativistic argument in superconductors *did* carry over to relativistic quantum fields. *The universe could behave like a superconductor after all.*

When Higgs wrote up his result and submitted it to the European journal *Physics Letters*, the paper was promptly rejected. The referee simply didn't think it was relevant to particle physics. So, Higgs added some passages commenting on possible observable consequences of his idea and submitted it to the US journal *Physical Review Letters*. In particular, he added, "It is worth noting that an essential feature of this type of theory is the prediction of incomplete multiplets of scalar and vector bosons."

In English this means that Higgs demonstrated that while one *could* remove the massless scalar particle (aka Goldstone boson) in favor of a massive vector particle (massive photon) in his model, there would also exist a leftover massive scalar (i.e., spinless) boson particle associated with the field whose condensate broke the symmetry in the first place. The Higgs boson was born.

Physical Review Letters promptly accepted the paper, but the referee asked Higgs to comment on the relation of his paper to a paper by François Englert and Robert Brout that had been received by the journal a month or so earlier. Much to Higgs's surprise, they had independently arrived at essentially the same conclusions. Indeed, the similarity between the papers is made clear by their titles. Higgs's paper was called "Broken Symmetries and the Masses of Gauge Bosons." The Englert and Brout paper was entitled "Broken Symmetry and the Mass of Gauge Vector Mesons." It is hard to imagine a closer match without coordinating names.

As if to add to the remarkable serendipity, twenty years later Higgs met Nambu at a conference and learned that Nambu had refereed both papers. How much more fitting could it be that the man who first brought the ideas of symmetry breaking and superconductivity to particle physics should referee the papers of the people who would demonstrate just how prescient this idea was. And like Nambu, all of these authors were fixated on the strong interaction, and on the possibility of figuring out how protons, neutrons, and mesons could have large masses.

Illustrating that the time was ripe for this discovery, within a month or so another team, Gerald Guralnik, C. R. Hagen, and Tom Kibble, also published a paper including many of the same ideas.

You may wonder why we call it the Higgs boson and not the Higgs-Brout-Englert-Guralnik-Hagen-Kibble boson. Besides the obvious answer that this label doesn't trip lightly off the tongue, of all the papers the only one to explicitly predict an accompanying massive scalar boson in massive gauge theories with spontaneous symmetry breaking was Higgs's paper. And, interestingly, Higgs only included the extra remark because the original version of his paper without that remark had been rejected.

One last bit of poetry. A couple of years after the original paper was published, Higgs completed a longer paper and was invited (in 1966) to speak at several locations in the USA, where he was spending a sabbatical year. After Higgs's talk at Harvard, where Sheldon Glashow was now a professor, Glashow apparently complimented him on having invented a "nice model" and moved on. Such was the fixation on the strong interaction that Glashow didn't realize then that this might be the key to resolving the issues in the weak interaction theory he had published five years earlier.

Part Three

REVELATION

Chapter 17

THE WRONG PLACE AT THE RIGHT TIME

Be not deceived: evil communications corrupt
good manners.

—1 CORINTHIANS 15:33

All of the six authors of the papers that describe what is most commonly called the Higgs mechanism (though after the recent Nobel Prize that Higgs shared with Englert, some are now calling it the BEH mechanism, for Brout, Englert, and Higgs) suspected and hoped that their work would help in understanding the strong force in nuclei. In their papers, any discussions of possible experimental probes of their ideas referenced the strong interaction—and in particular Sakurai's proposal of heavy vector mesons mediating this force. They hoped that a theory of the strong interaction that explained nuclear masses and short-range strong nuclear forces was around the corner.

Besides the general fascination with the strong nuclear force in nuclear physics, I suspect physicists tried to apply their new ideas to this theory for another reason. Given the range and strength of this force, the masses of new Yang-Mills-like particles that would be necessary to mediate the strong interaction would be comparable to the masses of protons and neutrons themselves and also of the other new particles being discovered in accelerators. Since experimental confirmation is the highest honor that theorists can achieve, it was natural to focus on understanding physics at these accessible energy scales, where new ideas, and new particles, could be quickly tested and explored in existing machines—with fame, if not fortune, around the corner. By contrast, as Schwinger had shown, any theory involving new particles associated with the weak force would require them to have masses several orders of magnitude larger than those available at accelerators at the time. This was clearly a problem to be considered at a later time, or so most physicists thought.

One of the many people who were fascinated by the physics of the strong interaction was the young theorist Steven Weinberg. There is poetry here as well. Weinberg grew up in New York City and attended the Bronx High School of Science, from which he graduated in 1950. One of his high school classmates was Sheldon Glashow, and the two of them moved together to study at Cornell University, living together in a temporary dorm there in their first semester before going their separate ways. While Glashow went to Harvard for graduate school, Weinberg moved on to Copenhagen—where Glashow would spend time as a postdoc—before arriving at Princeton to complete his PhD. Both of them were on the faculty at Berkeley in the early 1960s, leaving in the same year, 1966, for Harvard, where Glashow took up a professorship and Weinberg took a visiting position while on leave from Berkeley. Weinberg then moved to MIT in 1967, only to return to Harvard in 1973 to take the same chair and office that had been vacated by Julian Schwinger, Glashow's former supervisor. (When Weinberg moved into the office, he found in the closet a pair of shoes that Schwinger had left, clearly as a challenge to the younger scientist to try to fill them. He did.) When Weinberg left Harvard in 1982, Glashow then moved to occupy the same chair and office, but no shoes were left in the closet.

The lives of these two scientists were intertwined perhaps as closely as those of any other scientists in recent times, yet they form an interesting contrast. Glashow's brilliance is combined with an almost childlike enthusiasm for science. His strength lies in his creativity and his understanding of the experimental landscape and not so much in his detailed calculational abilities. By contrast, Weinberg is perhaps the most scholarly and serious (about physics) physicist I have ever known. While he has a wonderful ironic sense of humor, he never undertakes any physics project lightly, without the intent of mastering the relevant field. His physics textbooks are masterpieces, and his popular writing is lucid and full of wisdom. An avid reader of ancient history, Weinberg fully communicates the historical perspective not only on what he is doing, but on the whole physics enterprise.

Weinberg's approach to physics is like that of a steamroller. When I was at Harvard, we postdocs used to call Weinberg "Big Steve." When he was working on a problem, the best thing you could do was get out of the way, or you would be rolled over by the immense power of his intellect and energy. Earlier, before I moved to Harvard and was still at MIT, a friend of mine at the time, Lawrence Hall, was a graduate student at Harvard. Lawrence was ahead of me in his work, graduating before me. He told me that he was only able to complete the work that became his thesis with Weinberg because Weinberg had just won the Nobel Prize in 1979, and the ensuing hubbub forced him to slow down enough so that Lawrence could complete his calculations before Weinberg beat him to the punch.

One of the great fortunes of my life was to have the opportunity to work closely with both Glashow and Weinberg during the early and formative years of my own career. After Glashow helped rescue me from the black hole of mathematical physics, he became my collaborator at Harvard and for years later. Weinberg taught me much of what I know about particle theory. At MIT one doesn't have to take courses, just pass exams, so I only took one or two physics courses at MIT while working toward my PhD. But one of the perks of being at MIT was that I could take classes at Harvard. I took or sat in on every graduate class that

Weinberg taught during my graduate career, from quantum field theory onward. Glashow and Weinberg formed complementary role models for my own career. At my best I've tried to emulate aspects I learned from each of them, while recognizing that most often my "best" wasn't much in comparison.

Weinberg had, and has, a broad and abiding interest in the details of quantum field theory, and like many others during the early 1960s, he tried to focus on how one might understand the nature of the strong interaction using ideas of symmetry that, in large part due to the work of Gell-Mann, so dominated the field at the time.

Weinberg too was thinking about the possible application of ideas of symmetry breaking to understanding nuclear masses, based on Nambu's work, and like Higgs, Weinberg was quite disappointed by Goldstone's result that massless particles would always accompany such physics. So Weinberg decided, as he almost always did when he was interested in some physics idea, that he needed to prove it to himself. Thus his subsequent paper with Goldstone and Salam provided several independent proofs of the theorem in the context of strongly interacting particles and fields. Weinberg was so despondent about possible explanations of the strong interaction using spontaneous symmetry breaking that he added an epigraph to the draft of the paper that echoed Lear's response to Cordelia: "Nothing will come of nothing: speak again." (My book *A Universe from Nothing* makes plain why I am not a big fan of this quote. Quantum mechanics blurs the distinction between something and nothing.)

Weinberg subsequently learned about Higgs's (and others') result that one could get rid of unwanted massless Goldstone bosons that occur through symmetry breaking if the symmetry being broken was a gauge symmetry—where in this case the massless Goldstone bosons would disappear and otherwise massless gauge bosons would become massive—but Weinberg wasn't particularly impressed, viewing it as many other physicists did, as an interesting technicality.

Moreover, in the early 1960s the idea that the pion resembled in many

ways a Goldstone boson was useful in deriving some approximate for-
mulas for certain strong interaction reaction rates. Thus, the notion of
getting rid of Goldstone bosons in the strong interaction became less
attractive. Weinberg spent several years during this period exploring
these ideas. He worked out a theory whereby some symmetries that were
thought to be associated with the strong interaction might become broken spontaneously, and various strongly interacting vector gauge particles that convey the strong interaction might get masses via the Higgs
mechanism. The problem was he couldn't get agreement with observations without spoiling the initial gauge symmetry that would protect the
theory. The only way he could avoid this and preserve the initial gauge
symmetry he needed was if some vector particles became massive, and
others remained massless. But this disagreed with experiment.

Then one day in 1967 while driving in to MIT, he saw the light, literally and metaphorically. (I have driven with Steve in Boston, and while
I have lived to talk about it, I have seen how when he is thinking about
physics, all awareness of large masses such as other cars disappears.)
Weinberg suddenly realized that maybe he, and *everyone else*, was applying the right ideas of spontaneous symmetry breaking, but to the
wrong problem! Another example in nature could involve two different
vector bosons, one type massless and one type massive. The massless
vector boson could be the photon, and the massive one (or ones) could
be the massive mediator(s) of the weak interaction that had been speculated by Schwinger a decade earlier.

If this was true, then the weak and electromagnetic interactions
could be described by a unified set of gauge theories—one corresponding to the electromagnetic interaction (remaining unbroken) and one
corresponding to the weak interaction, with a broken-gauge symmetry
resulting in several massive mediators for that interaction.

In this case the world we live in would be *precisely* like a superconductor.

The weak interaction would be weak because of the simple accident
that the ground state of fields in our current universe breaks the gauge

symmetry that would otherwise govern the weak interaction symmetry. The photonlike gauge particles would get large masses, and as Schwinger had expected, the weak interaction would become so short-range that it would almost die off even on the length scale of protons and neutrons. This would also explain why neutron decay would happen so slowly.

The massive particles mediating the weak interaction would appear to us just as photons would appear to hypothetical physicists living inside a superconductor. So too the distinction between electromagnetism and the weak interaction would be just as illusory as the distinction that physicists on the ice crystals on that windowpane would make between forces along the direction of their icicle versus those perpendicular to that direction. It would be a simple accident that one gauge symmetry gets broken in the world of our experience, and the other doesn't.

Weinberg wanted to avoid thinking about strongly interacting particles since the situation there was still confused. So he decided to think about particles that interact only via the weak or electromagnetic interaction, namely electrons and neutrinos. Since the weak interaction turns electrons into neutrinos, he had to imagine a set of charged vector photonlike particles that would produce such a transformation. These are nothing other than the charged vector bosons that Schwinger envisaged, conventionally called W plus and W minus bosons.

Since only left-handed electrons and neutrinos get mixed together by the weak interaction, one type of gauge symmetry would have to govern just the interactions of left-handed particles with the W particles. But since both left-handed electrons *and* right-handed electrons interact with photons, the gauge symmetry of electromagnetism would somehow have to be incorporated in this unified model in such a way that left-handed electrons could interact with *both* photons *and* the new charged W bosons—while right-handed electrons would interact *only* with photons and not the W particles.

Mathematically, the only way to do this—as Sheldon Glashow had discovered when he was thinking about electroweak unification six years earlier—was if there was one additional *neutral* weak boson that right- and left-handed electrons could interact with in addition to interacting with photons. This new boson Weinberg dubbed the Z, zero.

A new field would have to exist in nature that would form a condensate in empty space to spontaneously break the symmetries governing the weak interaction. The elementary particle associated with this field would be the massive Higgs, while the remaining would-be Goldstone bosons would now be eaten by the W and Z bosons to make them massive, by the mechanism that Higgs first proposed. This would leave only the photon left over as a massless gauge boson.

But there's more. By virtue of the gauge symmetry he introduced, Weinberg's new Higgs particle would also interact with electrons, and when the condensate formed, the effect would be to give electrons a mass as well as the W and Z particles. Thus, not only would this model explain the masses of the gauge particles that mediate the weak force— and therefore determine the strength of that force—but the *same* Higgs field would also give electrons mass.

All the ingredients necessary for the unification of the weak and electromagnetic interaction were present in this model. Moreover, by starting with a Yang-Mills gauge theory with massless gauge bosons before symmetry breaking, there was hope that the same remarkable symmetry properties of gauge theories first exploited in quantum electrodynamics might also allow this theory to produce finite sensible results. While a fundamental theory with massive photonlike particles clearly had pathologies, the hope was that if the masses only resulted after symmetry breaking, these pathologies might not appear. But it was just a hope at the time.

Clearly in a realistic model the Higgs particle would couple to other particles engaged in the weak interaction, beyond the electron. In the absence of a Higgs condensate all these particles, protons, or the particles that made them up, and muons, etc., all of them would be exactly massless. *Every facet that is responsible for our existence*, indeed the very existence of the massive particles from which we are made, would thus arise as an accident of nature—the formation of a specific Higgs condensate in our universe. The particular features that make our world what it is—the galaxies, stars, planets, people, and the interactions among all of these—would be quite different if the condensate had never formed.

Or if it had formed differently.

Just as the world experienced by imaginary physicists on the ice crystal on that windowpane on a cold winter morning would have been completely different if the crystal had lined up in a different direction, so too the features of our world that allow our existence depend crucially on the nature of the Higgs condensate. What might seem so special about the features of the particles and fields that make up the world we live in would thus be no more special, planned, or significant than would be the accidental orientation of the spine of that ice crystal, even if it might appear to have special significance to beings living on the crystal.

And one last bit of poetry. The unique Yang-Mills model that Weinberg was driven to in 1967, which Abdus Salam would also stumble upon a year later, was precisely the model proposed six years earlier by his old high school friend Sheldon Glashow when he responded to Schwinger's challenge to find a symmetry that might unify the weak and electromagnetic interactions. No other choice could mathematically reproduce what we see in the world today. Glashow's model had been largely ignored in the interim because no mechanism was then known to give the weak bosons masses. But now such a mechanism existed, the Higgs mechanism.

Weinberg and Glashow, whose lives had crisscrossed since they were children, would later share the Nobel Prize, along with Salam, for *completely independent* discoveries of the greatest unification in physical theory since Maxwell had unified electricity and magnetism and Einstein had unified space and time.

Chapter 18

THE FOG LIFTS

*Their voice goes out through all the earth, and their
words to the end of the world.*

—PSALM 19:4

You might expect that physicists around the world would
have thrown parties with fireworks when Weinberg's paper came out.
But for the next three years following publication of Weinberg's theory,
not a single physicist, not even Weinberg himself, would find cause to
reference the paper—now one of the most highly cited papers in all of
particle physics. If a great discovery about nature had been made, no
one had yet noticed.

After all, Maxwell's unification made the beautiful prediction that
light was an electromagnetic wave whose speed could be calculated from
first principles, and lo and behold, the prediction was equal to the measured speed of light. Einstein's unification of space and time predicted
that clocks would slow for moving observers, and lo and behold, they do,
and in just the way he predicted. In 1967 the Glashow-Weinberg-Salam
unification of the weak and electromagnetic interactions predicted
three new vector bosons that were almost one hundred times heavier
than any particle that had been yet detected. It also predicted new in-

teractions between electrons and neutrinos and matter due to the newly predicted Z particle that had not only not been seen, but a number of experiments suggested did not exist. It also required the existence of a new and as yet unobserved massive fundamental scalar boson, the Higgs particle, when no fundamental scalar particles were yet known to exist in nature. And finally, as a quantum theory, no one knew if it made sense.

Is it any wonder that the idea did not immediately catch fire? Nevertheless, within a decade everything would change, resulting in the most theoretically productive period for elementary particle physics since the discovery of quantum mechanics. While a gauge theory of the weak interaction started the ball rolling, what resulted was far greater.

· · ·

The first crack in the dike holding back the waters of progress came, fittingly, with the work of Dutch graduate student Gerardus 't Hooft, in 1971. I always remember how to spell his name because a particularly brilliant and witty former Harvard colleague, the late Sidney Coleman, used to say that if Gerard had monogrammed cuff links, they would need an apostrophe on them. Before 1971 many of the greatest theorists in the world had tried to figure out whether the infinities that plague most quantum field theories would disappear for spontaneously broken gauge theories as they do for their unbroken cousins. But the answer eluded them. Remarkably this young graduate student, working under the supervision of a seasoned pro—Martinus Veltman—found a proof that others had missed. Often when presented with a new result, we physicists can work through the details and imagine how we might have discovered it ourselves. But many of 't Hooft's insights, and there were many—almost all the new ideas in the 1970s derived in one way or another from his theoretical inventions—seemed to come from some hidden reservoir of intuition.

The other remarkable thing about Gerard is how gentle, shy, and unassuming he is. For someone who became famous in the field when he was a student, one might have expected some sense of privilege. But from the first time I met him—again when I was a lowly graduate student—Gerard treated me as an interesting friend, and I am pleased to say that relationship has continued. I always try to remember this attitude when I meet young students who may seem shy or intimidated, and I try to emulate Gerard's open generosity of spirit.

His supervisor Tini Veltman, as he is often called, couldn't appear more different. Not that Tini isn't fun to talk to. He is. But he always made explicitly clear to me the moment we started a discussion that whatever I might say, I didn't understand things well enough. I always enjoyed the challenge.

It is important to note that 't Hooft would never have approached the problem if Veltman had not been obsessed with it, even as most others gave up. The notion that one might ultimately extend the techniques that Feynman and others had developed to tame quantum electrodynamics to try to understand more complex theories such as spontaneously broken Yang-Mills theory was simply viewed as naïve by many in the field. But Veltman stayed with the project, and he wisely found a graduate student who was also a genius to help him.

It took a while for 't Hooft's and Veltman's ideas to sink in and the new techniques 't Hooft had developed to become universally adopted, but within a year or so physicists agreed that the theory that Weinberg, and later Salam, had proposed, made sense. Citations of Weinberg's paper suddenly began to grow exponentially. But making sense and being right are two different things. Did nature actually use the specific theory that Glashow, Weinberg, and Salam had suggested?

That remained the key open question, and for a while it looked as if the answer was no.

The existence of the new neutral particle, the Z, required by the theory, was a significant addition, beyond the charged particles suggested years earlier by Schwinger and others that were required to change neutrons into protons and electrons into neutrinos. It meant that there would be a new kind of weak interaction, not just for electrons and neutrinos but also for protons and neutrons, mediated by a new neutral-particle exchange. In this case, as for electromagnetism, the identity of the particles interacting would not change. Such interactions became known as neutral current interactions, and the obvious way to test the theory was to look for them. The best place to look for them was in the interactions of the only particles in nature that just feel the weak interaction, namely neutrinos.

You may recall that the prediction of such neutral currents was one of the reasons that Glashow's 1961 suggestion never caught on. But Glashow's model wasn't a full theory. Particle masses were simply put into the equations by hand, and as a result quantum corrections couldn't be controlled. However, when Weinberg and Salam proposed their model for electroweak unification, all elements that allowed for detailed predictions were there. The mass of the Z particle was predicted, and as 't Hooft had shown, one could calculate all quantum corrections in a reliable way, just as one did for quantum electrodynamics.

This was a good thing, and a bad thing because no wiggle room was left to argue away any possible disagreements with observation. And in 1967 there appeared to be such disagreements. No such neutral currents had been observed in high-energy collisions of neutrinos with protons, with an upper limit being set of about 10 percent of the rate observed for more familiar charge-changing weak interactions of neutrinos and protons, such as neutron decay. Things looked bad, and most physicists assumed weak neutral currents didn't exist.

Weinberg had a vested interest in this quest, and in 1971 he reasonably argued that there was still wiggle room. But this view was not generally held by others in the community.

In the early 1970s, new experiments at the European Organization for Nuclear Research (CERN) in Geneva were performed using the proton accelerator there, which smashed high-energy protons into a long target. Most particles produced in the collision would be absorbed in the target, but neutrinos would emerge from the other end—as their interactions are so weak that they could traverse the target without being absorbed. The resulting high-energy neutrino beam would then strike a detector placed in its path that could record the few events in which neutrinos might interact with the detector material.

A huge new detector was built, named Gargamelle after the giantess mother of Gargantua, from the work of the French writer Rabelais. This five-meter-by-two-meter "bubble chamber" vessel was filled with a superheated liquid in which trails of bubbles would form when an energetic charged particle traversed it, sort of like seeing the vapor trail high in the sky of a plane that is itself not visible.

Interestingly, when the experimentalists who built Gargamelle met in 1968 to discuss their plans for neutrino experiments, the idea of searching for neutral currents wasn't even mentioned—an indication of how many physicists thought the issue was then settled. Of far more interest to them was the possibility of following up on recent exciting experiments at the Stanford Linear Accelerator (SLAC), where high-energy electrons had been used as probes to explore the structure of protons. Using neutrinos as probes of protons might give cleaner measurements because the neutrinos are not charged.

After the results of 't Hooft and Veltman, however, in 1972, experimentalists began to take the gauge theory description of the weak interaction, and in particular the Glashow-Weinberg-Salam proposal, seriously. That meant looking for neutral currents. The Gargamelle collaboration had the capability to do this, in principle, even though it hadn't been designed for the task.

Most of the high-energy neutrinos in the beam would interact with protons in the target by turning into muons, the heavier partners of electrons. The muons would exit the target, producing a long charged-particle track all the way to the edge of the detector. The protons would be converted into neutrons, which would themselves not produce a track but would collide with nuclei, producing a short shower of charged particles that would leave tracks. Thus, the experiment was designed to detect muon tracks, as well as accompanying charged-particle showers, both arising as separate signals of a single weak interaction.

However, sometimes a neutrino would interact with material outside the detector, producing a neutron that might recoil back into the detector and then interact there. Such events would consist of a single strongly interacting shower of particles due to the colliding neutron, with no accompanying muon track.

When Gargamelle began to search for neutral current events, such isolated charged-particle showers without an accompanying muon became just the signal the scientists needed to focus on. In neutral current events a neutrino that interacts with a neutron or proton in the detector doesn't convert into a charged muon, but simply bounces off and escapes the detector unobserved. All that would be observable would be the recoiling nuclear shower—the same signature produced by the more standard neutrino interactions outside the detector that produce neutrons that recoil back into the detector and produce a nuclear shower.

The challenge, then, if the experiment was to definitively detect neutral current events, was to distinguish neutrino-induced events from such neutron-induced events. (This same problem has provided the chief challenge to experimentalists looking for any weakly interacting particles, including the presumed dark matter particles that are being searched for in underground detectors around the world today.)

The observation of a single recoil electron, with no other charged-particle tracks in the detector, was observed in early 1973. This could have arisen from the less frequent predicted neutral current collisions of neutrinos with electrons instead of protons or neutrons. But generally a single event is not enough to definitively claim a new discovery in particle physics. However, it did give hope, and by March of 1973 a careful analysis of neutron backgrounds and observed isolated particle showers appeared to provide evidence that weak neutral current interactions actually exist. Nevertheless, not until July of 1973 did the researchers at

CERN complete a sufficient number of checks to be confident enough to claim a detection of neutral currents, which they did at a conference in Bonn in August.

The story might have ended there, but unfortunately, shortly after this, another collaboration searching for neutral currents rechecked their apparatus and found that a previous signal for neutral currents had disappeared. This produced significant confusion and skepticism in the physics community, where once again neutral currents seemed suspect. Ultimately the Gargamelle collaboration returned to the drawing board, tested the detector using a proton beam directly, and took a great deal more data. At a conference almost a year later, in June 1974, the Gargamelle collaboration presented overwhelming confirmation of the signal. Meanwhile the competing collaboration had found the cause of its error and confirmed the Gargamelle result. Glashow, Weinberg, and Salam were vindicated.

Neutral currents had arrived, and a remarkable unification of the weak and the electromagnetic interactions appeared to be at hand. But two loose ends still remained to be cleared up.

The existence of neutral currents in neutrino scattering validated the notion that the Z particle existed, but this didn't guarantee that the weak interaction was identical to that proposed by Glashow, Weinberg, and Salam, where the weak and the electromagnetic interactions were unified. To explore this required an experiment using a particle that participated in both the weak and the electromagnetic interaction. The electron was ideal for this purpose because these are the only two interactions it experiences.

When electrons interact with other charges by their electromagnetic attraction, left-handed electrons and right-handed electrons behave identically. However, the Glashow-Weinberg-Salam theory required that weak interactions occur differently for left-handed versus right-handed particles. This implied that careful measurements of the scattering of polarized electrons—electrons prepared initially in left- or right-handed states using magnetic fields—off various targets should reveal a violation of left-right symmetry, but not as extreme an asymmetry as that observed in neutrino scattering—because the neutrino is purely left-handed. The degree of violation in electron scattering, if it existed, would then reflect the extent to which the weak interaction and electromagnetism were mixed together in a unified theory.

The idea of testing for such interference using electron scattering had actually been suggested as early as 1958 by the remarkable Soviet physicist Yakov B. Zel'dovich. But it would take twenty years for sufficiently sensitive experiments to actually take place. And as for the neutral current discovery, the road to success was full of potholes and wrong turns along the way.

One of the reasons it took so long to test this idea is that the weak interaction is weak. Because the dominant interaction of electrons with matter is electromagnetic, the left-right asymmetry predicted due to a possible exchange of a Z particle was small, smaller than one part in ten thousand. To test for such a small asymmetry required both an intense beam and one whose initial polarization was well determined.

The best place to perform these experiments was at the Stanford Linear Accelerator, a two-mile-long electron linear accelerator built in 1962 that was the longest and straightest structure that had ever been built. In 1970 polarized beams were introduced, but not until 1978 was an experiment designed and run with the sensitivity required to look for weak-electromagnetic interference in electron scattering.

While the successful observation of neutral currents in 1974 meant that the Glashow-Weinberg-Salam theory began to have wide acceptance among theorists, what made the 1978 SLAC experiment so important was that in 1977 two atomic physics experiments had reported results that, if correct, convincingly ruled out the theory.

In our story thus far, light has played a crucial role, illuminating (if you will forgive the pun) our understanding not only of electricity and magnetism, but space, time, and ultimately the nature of the quantum world. So too it was realized that light could help probe for a possible electroweak unification.

The first great success of quantum electrodynamics was the correct prediction of the spectrum of hydrogen, and eventually other atoms. But if electrons also feel the weak force, then this will provide a small additional force between electrons and nuclei that should alter—if slightly—the characteristics of their atomic orbits. For the most part these are unobservable because electromagnetic effects swamp weak effects. But weak interactions violate parity, so the same weak-electromagnetic neutral current interference that was being explored using polarized electron beams can produce novel effects in atoms that would vanish if electromagnetism was the only force involved.

In particular, for heavy atoms, the Glashow-Weinberg-Salam theory predicted that if polarized light was transmitted through a gas of atoms, then the direction of the polarization of the light would be rotated by about a millionth of a degree, due to parity-violating neutral current effects in the atoms through which the light passed.

In 1977 the results of two independent atomic physics experiments, in Seattle and Oxford, were published in back-to-back articles in *Physical Review Letters*. The results were dismaying. No such optical rotation was seen at a level ten times smaller than that predicted by the electroweak theory. Had only one experiment reported the result, it would have been more equivocal. But the same result from two independent experiments using independent techniques made it appear definitive. The theory appeared to be ruled out.

Nevertheless, the SLAC experiment, which had begun three years earlier, was well under way, and since all of the experimental preparation had begun, the experiment was approved to begin to take data in early 1978. Because of the earlier null results from the atomic physics experiments, the Stanford collaboration added several bells and whistles to the experiment so that if they saw no effect, they could guarantee that they *could* have seen such an effect were it there.

Within two months the experiment began to show clear signs of parity violation, and by June 1978 the scientists announced a nonzero result, in agreement with the predictions of the Glashow-Weinberg-Salam model, based on measured neutrino neutral current scattering, which measured the strength of the Z interaction.

Still, questions remained, especially given the apparent disagreement with the Seattle/Oxford results. At a talk at Caltech on the subject, Richard Feynman, characteristically, homed in on a key outstanding experimental question and asked whether the SLAC experimentalists had checked that the detector responded equally well to both left-handed and right-handed electrons. They hadn't, but for theoretical reasons they had had no reason to expect the detectors to behave differently for the different polarizations. (Feynman would famously get to the heart of another complex problem eight years later after the tragic *Challenger* explosion, when he simply demonstrated the failure of an O-ring seal to the investigating commission and to the public watching the televised proceedings.)

Over the fall the SLAC experiment refined their efforts to rule out both this concern and others that had been raised, and by the fall they reported a definitive result in agreement with the Glashow-Weinberg-Salam prediction, with an uncertainty of less than 10 percent. Electroweak unification was vindicated!

To date, I don't know if anyone has a good explanation of why the original atomic physics results were wrong (later experiments agreed with the Glashow-Weinberg-Salam theory) except that the experiments, and the theoretical interpretation of the experiments, are hard.

But a mere year later, in October 1979, Sheldon Glashow, Abdus Salam, and Steven Weinberg were awarded the Nobel Prize for their electroweak theory, now validated by experiment, that unified two of the four forces of nature based on a single fundamental symmetry, gauge invariance. If the gauge symmetry hadn't been broken, hidden from view, the weak and electromagnetic interactions would look identical. But then all of the particles that make us up wouldn't have mass, and we wouldn't be here to notice. . . .

This is not the end of our story, however. Two out of four is still only two out of four. The strong interaction, which had motivated much of the work that led to electroweak unification, had continued to stubbornly resist all attempts at explanation even as the electroweak theory took shape. No explanation of the strong nuclear force via spontaneously broken gauge symmetries met the test of experiment.

Thus, even as scientist-philosophers of the twentieth century had stumbled—often by a convoluted and dimly lit path—outside our cave of shadows to glimpse the otherwise hidden reality beneath the surface, one more force relevant to understanding the fundamental structure of matter was conspicuously missing from the beautiful emerging tapestry of nature.

Chapter 19

FREE AT LAST

Let my people go.

—EXODUS 9:1

The long road that led to electroweak unification was a tour de force of intellectual perseverance and ingenuity. But it was also a detour de force. Almost all of the major ideas introduced by Yang, Mills, Yukawa, Higgs, and others that led to this theory were developed in the apparently unsuccessful struggle to understand the strongest force in nature, the strong nuclear force. Recall that this force, and the strongly interacting particles that manifested it, had so bedeviled physicists that in the 1960s many of them had given up hope of ever explaining it via the techniques of quantum field theory that had so successfully now described both electromagnetism and the weak interaction.

There had been one success, centered on Gell-Mann and Zweig's proposal that all the strongly interacting particles that had been observed, including the proton and the neutron, could be understood as being made up of more fundamental objects, which, as I have described, Gell-Mann called quarks. All the known strongly interacting particles, and at the time undiscovered particles, could be classified assuming they were made of quarks. Moreover, the symmetry arguments that led

Gell-Mann in particular to come up with his model served as the basis for making some sense of the otherwise confusing data associated with the reactions of strongly interacting matter.

Nevertheless, Gell-Mann had allowed that his scheme might merely be a mathematical construct, useful for classification, and that quarks might not represent real particles. After all, no free quarks had ever been observed in accelerators or cosmic-ray experiments. He was also probably influenced by the popular idea that quantum field theory, and hence the notion of elementary particles themselves, broke down on nuclear scales. Even as late as 1972 Gell-Mann stated, "Let us end by emphasizing our main point, that it may well be possible to construct an explicit theory of hadrons, based on quarks and some kind of glue. . . . Since the entities we start with are fictitious, there is no need for any conflict with the bootstrap . . . point of view."

Viewed in this context, the effort to describe the strong interaction by a Yang-Mills gauge quantum field theory, with real gauge particles mediating the force, would be misplaced. It also seemed impossible. The strong force appeared to operate only on nuclear scales, so if it was to be described by a gauge theory, the photonlike particles that would convey the force would have to be heavy. But there was also no evidence of a Higgs mechanism, with massive strongly interacting Higgs-like particles, which experiments could have easily detected. Compounding this, the force was simply so strong that even if it was described by a gauge theory, then all of the quantum field theory techniques developed for deriving predictions—which worked so well for the other forces—would have broken down if applied to the strong force. This is why Gell-Mann in his quote referred to the "bootstrap"—the Zen-like idea that no particles were truly fundamental. The sound of no hands clapping, if you will.

Whenever theory faces an impasse like this, it sure helps to have experiment as a guide, and that is exactly what happened, in 1968. A series of pivotal experiments, performed by Henry Kendall, Jerry Friedman, and Richard Taylor, using the newly built SLAC accelerator to scatter high-energy electrons off protons and neutrons, revealed something remarkable. Protons and neutrons *did* appear to have some substructure, but it was strange. The collisions had properties no one had expected. Was the signal due to quarks?

Theorists were quick to come to the rescue. James Bjorken demonstrated that the phenomena observed by the experimentalists, called scaling, could be understood if protons and neutrons were composed of virtually noninteracting pointlike particles. Feynman then interpreted these objects as real particles, which he dubbed partons, and suggested they could be identified with Gell-Mann's quarks.

This picture had a *big* problem, however. If all strongly interacting particles were composed of quarks, then quarks should surely be strongly interacting themselves. Why should they appear to be almost free inside protons and neutrons and not be interacting strongly with each other?

Moreover, in 1965, Nambu, Moo-Young Han, and Oscar Greenberg had convincingly argued that, if strongly interacting particles were composed of quarks and if they were fermions, like electrons, then Gell-Mann's classification of known particles by various combinations of quarks would only be consistent if quarks possessed some new kind of internal charge, a new Yang-Mills gauge charge. This would imply that they interacted strongly via a new set of gauge bosons, which were then called gluons. But where were the gluons, and where were the quarks, and why was there no evidence of quarks interacting strongly inside protons and neutrons if they were really to be identified with Feynman's partons?

In yet another problem with quarks, protons and neutrons have weak interactions, and if these particles were made up of quarks, then the quarks would also have to have weak interactions in addition to strong interactions. Gell-Mann had identified three different types of quarks as comprising all known strongly interacting particles at the time. Mesons could be comprised of quark-antiquark pairs. Protons and neutrons could be made up of three fractionally charged quarks, which Gell-Mann called up (u) and down (d) quarks. The proton would be made of two up quarks and one down quark, while the neutron would be made of two down quarks and one up quark. In addition to these two types of quarks, one additional type of quark, a heavier version of the down quark, was required to make up exotic new elementary particles. Gell-Mann called this the strange (s) quark, and particles containing s quarks were dubbed to possess "strangeness."

When neutral currents were first proposed as part of the weak interaction, this created a problem. If quarks interacted with the Z particles, then u, d, and s quarks could remain u, d, and s quarks before and after the neutral current interaction, just as electrons remained electrons before and after the interaction. However, because the d and s quarks had precisely the same electric and isotopic spin charges, nothing would prevent an s quark from converting into a d quark when it interacted with a Z particle. This would allow particles containing s quarks to decay into particles containing d quarks. But no such "strangeness-changing decays" were observed, with high sensitivity in experiments. Something was wrong.

This absence of "strangeness-changing neutral currents" was explained brilliantly, at least in principle, by Sheldon Glashow, along with collaborators John Iliopoulos and Luciano Maiani, in 1970. They took the quark model seriously and suggested that if a fourth quark, dubbed a charm (c) quark, existed, which had the same charge as the u quark, then a remarkable mathematical cancellation could occur in the calculated transformation rate for an s quark into a d quark, and strangeness-changing neutral currents would be suppressed, in agreement with experiments.

Moreover, this scheme began to suggest a nice symmetry between quarks and particles such as electrons and muons, all of which could exist in pairs associated with the weak force. The electron would be paired with its own neutrino, as would the muon. The up and down quarks would form one pair, and the charm and the strange quark another pair. W particles interacting with one particle in each pair would turn it into the other particle in the pair.

None of these arguments addressed the central problems of the strong interaction between quarks, however. Why had no one ever observed a quark? And, if the strong interaction was described by a gauge theory with gluons as the gauge particles, how come no one had ever observed a gluon? And if the gluons were massless, how come the strong force was short-range?

These problems continued to suggest to some that quantum field theory was the wrong approach for understanding the strong force. Freeman Dyson, who had played such an important role in the development of the first successful quantum field theory, quantum electrodynamics, asserted, when describing the strong interaction, "The correct theory will not be found in the next hundred years."

One of those who were convinced that quantum field theory was doomed was a brilliant young theorist, David Gross. Trained under Geoffrey Chew, the inventor of the bootstrap picture of nuclear democracy, in which elementary particles were an illusion masking a structure in which only symmetries and not particles were real, Gross was well primed to try to kill quantum field theory for good.

Recall that even as late as 1965, when Richard Feynman received his Nobel Prize, it was still felt that the procedure he and others had developed for getting rid of infinities in quantum field theory was a trick—that something was fundamentally wrong at small scales with the picture that quantum field theory presented.

Russian physicist Lev Landau had shown in the 1950s that the electric charge on an electron depends on the scale at which you measure it. Virtual particles pop out of empty space, and electrons and all other elementary particles are surrounded by a cloud of virtual particle-antiparticle pairs. These pairs screen the charge, just as a charge in a dielectric material gets screened. Positively charged virtual particles tend to closely surround the negative charge, and so at a distance the physical effects of the initial negative charge are reduced.

This meant, according to Landau, that the closer you get to an electron, the larger its actual charge will appear. If we measure the electron charge to be some specific value at large distances, as we do, that would mean that the "bare" charge on the electron—namely the charge on the fundamental particle considered without all the infinite dressing by particle-antiparticle pairs surrounding it on ever-smaller scales—would have to be infinite. Clearly something was rotten with this picture.

Gross was influenced not only by his supervisor, but also by the prevailing sentiments of the time, mostly arguments by Gell-Mann, who dominated theoretical particle physics in the late fifties and early sixties. Gell-Mann advocated using algebraic relations that arise from thinking about field theories, then keeping the relations and throwing away the field theory. In a particularly Gell-Mann-esque description, he stated, "We may compare this process to a method sometimes employed in French cuisine: a piece of pheasant meat is cooked between two slices of veal, which are then discarded."

Thus one could abstract out properties of quarks that might be useful for predictions, but then ignore the actual possible existence of quarks. However, Gross began to be disenchanted by just using ideas associated with global symmetries and algebras and longed to explore dynamics that might actually describe the physical processes that were occurring inside strongly interacting particles. Gross and his collaborator Curtis Callan built upon earlier work by James Bjorken to show that the charged particle apparently located inside protons and neutrons had to have spin ½, identical to that of electrons. Later, with other collaborators, Gross showed that a similar analysis of neutrino scattering off protons and neutrons as measured at CERN revealed that the components looked just like the quarks that Gell-Mann had proposed.

If it quacks like a duck and walks like a duck, it is probably a duck. Thus, for Gross, and others, the reality of quarks was now convincing.

But as convinced as many such as Gross were by the reality of quarks, they were equally convinced that this implied that field theory could not possibly be the correct way to describe the strong interaction. The results of the experiment required the constituents to be essentially non-interacting, not strongly interacting.

In 1969 Gross's colleagues at Princeton Curtis Callan and Kurt Symanzik rediscovered a set of equations explored by Landau, and then Gell-Mann and Francis Low, that described how quantities in quantum field theory might evolve with scale. If the partons inferred by the SLAC experiments had any interactions at all—as quarks must have—then measurable departures from the scaling that Bjorken had derived would occur, and the results that Gross and his collaborators had also derived when comparing theory and the SLAC experiments would also have to be modified.

Over the next two years, with the results of 't Hooft and Veltman, and the growing success of the predictions of the theory of the weak and electromagnetic interactions, more people began to turn their attention once again to quantum field theory. Gross decided to prove in great generality that no sensible quantum field theory could possibly reproduce the experimental results about the nature of protons and neutrons observed at SLAC. Thus he hoped to kill this whole approach to attempting to understand the strong interaction. First, he would prove that the only way to explain the SLAC results was if somehow, at short distances, the strength of the quantum field interactions would have to go to zero, i.e., the fields would essentially become noninteracting at short distances. Then, after that, he would show that no quantum field theory had this property.

Recall that Landau had shown that quantum electrodynamics, the prototypical consistent quantum field theory, has precisely the opposite behavior. The strength of electric charges becomes larger as the scale at which you probe particles (such as electrons) gets smaller due to the cloud of virtual particles and antiparticles surrounding them.

Early in 1973 Gross and his collaborator Giorgio Parisi had completed the first part of the proof, namely that scaling as observed at SLAC implied the strong interactions of the proton's constituents must go to zero at small-distance scales if the strong nuclear force was to be described by any fundamental quantum field theory.

Next, Gross attempted to show that no field theories actually had this behavior—the strength of interactions going to zero at small-distance scales—which he dubbed asymptotic freedom. With help from Harvard's Sidney Coleman, who was visiting Princeton at the time, Gross was able to complete this proof for all sensible quantum field theories, except for Yang-Mills-type gauge theories.

Gross now took on a new graduate student, twenty-one-year-old Frank Wilczek, who had come to Princeton from the University of Chicago planning to study mathematics, but who switched to physics after taking Gross's graduate class in field theory.

Gross was either lucky or astute because he served as the graduate supervisor of probably the two most remarkable intellects among physicists in my generation, Wilczek and Edward Witten, who helped lead the string theory revolution in the 1980s and '90s and who is the only physicist ever to win the prestigious Fields Medal, the highest award given to mathematicians. Wilczek is probably one of the few true physics polymaths. Frank and I became frequent collaborators and friends in the early 1980s, and he is not only one of the most creative physicists I have ever worked with, he also has an encyclopedic knowledge of the field. He has read almost every physics text ever written, and he has assimilated the information. In the intervening years, he has made numerous fundamental contributions not only to particle physics, but to cosmology and also the physics of materials.

Gross assigned Wilczek to explore with him the one remaining loophole in Gross's previous proof—determining how the strength of the interaction in Yang-Mills theories changed as one went to shorter-distance scales—to prove that these theories too could not exhibit asymptotic freedom. They decided to directly and explicitly calculate the behavior of the interactions in the theories at shorter and shorter-distance scales.

This was a formidable task. Since that time tools have been developed for doing the calculation as a homework problem in a graduate course. Moreover, things are always easier to calculate when you know what the answer will be, as we now do. After several hectic months, with numerous false starts and numerical errors, in February of 1973 they completed their calculations and discovered, to Gross's great surprise, that in fact Yang-Mills theories *are* asymptotically free—the interaction strength in these theories does approach zero as interacting particles get closer together. As Gross later put it, in his Nobel address, "For me the discovery of asymptotic freedom was totally unexpected. Like an atheist who has just received a message from a burning bush, I became an immediate true believer."

Sidney Coleman had assigned his own graduate student David Politzer to do a similar calculation, and his independent result agreed with Gross and Wilczek's and was obtained at about the same time. That the results agreed gave both groups greater confidence in them.

Not only can Yang-Mills theories be asymptotically free, they are the *only* field theories that are. This led Gross and Wilczek to suggest, in the opening of their landmark paper, that because of this uniqueness, and because asymptotic freedom seemed to be required for any theory of the strong interaction given the 1968 SLAC experimental results, perhaps a Yang-Mills theory *could* explain the strong interaction.

Which Yang-Mills theory was the right one needed to be determined, and also why the massless gauge particles that are the hallmark of Yang-Mills theories had not been seen. And related to this, perhaps the most important long-standing question remained: Where were the quarks?

But before I address these questions, you might be wondering why Yang-Mills theories have such a different behavior from their simpler cousin quantum electrodynamics, where Landau had shown the strength of the interaction between electric charges gets larger on small-distance scales.

The key is somewhat subtle and lies in the nature of the massless gauge particles in Yang-Mills theory. Unlike photons in QED, which have no electric charge, the gluons that were predicted to mediate the strong interaction possess Yang-Mills charges, and therefore gluons interact with each other. But because Yang-Mills theories are more complicated than QED, the charges on gluons are also more complicated than the simple electric charges on electrons. Each gluon not only looks like a charged particle, but also like a little charged magnet.

If you bring a small magnet near some iron, the iron gets magnetized and you end up with a more powerful magnet. Something similar happens with Yang-Mills theories. If I have some particle with a Yang-Mills charge, say, a quark, then quarks and antiquarks can pop out of the vacuum around the charge and screen it, as happens in electromagnetism. But gluons can also pop out of the vacuum, and since they act like little magnets, they tend to align themselves along the direction of the field produced by the original quark. This increases the strength of the field, which in turn induces more gluons to pop out of the vacuum, which further increases the field, and so on.

As a result, the deeper into the virtual gluon cloud you penetrate—i.e., the closer you get to the quark—the weaker the field will look. Ultimately, as you bring two quarks closer together, the interaction will get so weak that they will begin to act as if they are not interacting at all, the characteristic of asymptotic freedom.

I used gluons and quarks as labels here, but the discovery of asymptotic freedom did not point uniquely to any specific Yang-Mills theory. However, Gross and Wilczek recognized the natural candidate was the Yang-Mills theory that Greenberg and others had posited was necessary for Gell-Mann's quark hypothesis to explain the observed nature of elementary particles. In this theory each quark carries one of three different types of charges, which are labeled, for lack of better names, by colors, say, red, green, or blue. Because of this nomenclature Gell-Mann coined a name for this Yang-Mills theory: quantum chromodynamics (QCD), the quantum theory of colored charges, in analogy to quantum electrodynamics, the quantum theory of electric charges.

Gross and Wilczek posited, based on the observational arguments in favor of such a symmetry associated with quarks, that quantum chromodynamics was the *correct* gauge theory of the strong interaction of quarks.

The remarkable idea of asymptotic freedom got an equally remarkable experimental boost within a year or so of these theoretical developments. Experiments at SLAC and at another accelerator in Brookhaven, Long Island, made the striking and unexpected discovery of a new massive elementary particle that appeared as if it might be made up of a new quark—indeed, the so-called charmed quark that had been predicted by Glashow and friends four years earlier.

But this new discovery was peculiar, because the new particle lived far longer than one might imagine based on the measured lifetime of unstable lighter strongly interacting particles. As the experimentalists who discovered this new particle said, observing it was like wandering in the jungle and finding a new species of humans who lived not up to one hundred, but up to ten thousand years.

Had the discovery been made even five years earlier, it would have seemed inexplicable. But in this case, fortune favored the prepared mind. Tom Appelquist and David Politzer, both at Harvard at the time, quickly realized that if asymptotic freedom was indeed a property of the strong interaction, then one could show that the interactions governing more massive quarks would be less strong than the interactions governing the lighter, more familiar quarks. Interactions that are less strong would mean particles decay less quickly. What would otherwise have been a mystery was in this case a verification of the new idea of asymptotic freedom. Everything seemed to be fitting into place.

Except for one pretty *big* thing. If the theory of quantum chromodynamics was a theory of the interactions of quarks and gluons, where *were* the quarks and gluons? How come none had ever been seen in an experiment?

Asymptotic freedom provides a key clue. If the strength of the strong interaction gets weaker the closer one gets to a quark, then conversely it should get stronger and stronger the farther one is away from the quark. Imagine, then, what happens if I have a quark and an antiquark that are bound together by the strong interaction and I try to pull them apart. As I try to pull them apart, I need more and more energy because the strength of the attraction between them grows with distance. Eventually so much energy becomes stored in the fields surrounding the quarks that it becomes energetically favorable instead for a new quark-antiquark pair to pop out of the vacuum and then for each to become bound to one of the original particles. The process is shown schematically below.

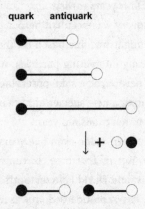

It would be like stretching a rubber band. Eventually the band will snap into two pieces instead of stretching forever. Each piece in this case would then represent a new bound quark-antiquark pair.

What would this mean for experiments? Well, if I accelerate a particle such as an electron and it collides with a quark inside a proton, it will kick the quark out of the proton. But as the quark begins to exit the proton, the interactions of the quark with the remaining quarks will increase, and it will eventually be energetically favored for virtual quark-antiquark pairs to pop out of the vacuum and bind to both the ejected quark and the other quarks as well. This means that one will create a shower of strongly interacting particles, such as protons or neutrons or pions or so on, moving along the direction of the original ejected quark, and similarly a shower of strongly interacting particles recoiling in the direction of motion of the original remaining quarks left over from the proton. One will never see the quarks themselves.

Similarly, if a particle collides with a quark, in recoiling sometimes the quark will emit a gluon before it binds with an antiquark popping out of the vacuum. Then since gluons interact with each other as well as with quarks, the new gluon might emit more gluons. The gluons in turn will be surrounded by new quarks that pop out of the vacuum, creating new strongly interacting particles moving along the direction of each original gluon. In this case one would expect in some cases to see not a single shower moving in the direction of the original quark, but several showers, corresponding to each new gluon that is emitted along the way.

Because quantum chromodynamics is a specific, well-defined theory, one can predict the rate at which quarks will emit gluons, and the rate at which one would see a single shower, or jet as it is called, kicked out when an electron collides with a proton or neutron, and the rate at which one would see two showers, and so on. Eventually, when accelerators became powerful enough to observe all these processes, the observed rates agreed well with the predictions of the theory.

There is every reason to believe that this picture of free quarks and gluons quickly getting bound to new quarks and antiquarks so that one would never observe a free quark or gluon is valid. This is called *Confinement* because quarks and gluons are always confined inside strongly interacting particles such as protons and neutrons and can never break free from them without getting confined in newly created strongly interacting particles.

Since the actual process by which the quarks get confined occurs as the forces become stronger and stronger when the quark moves farther and farther away from its original companions, the standard calculations of quantum field theory, which are valid when the interactions are not too strong, break down. So this picture, validated by experiment, cannot be fully confirmed by tractable calculations at the moment.

Will we ever derive the necessary mathematical tools to analytically demonstrate from first principles that confinement is indeed a mathematical property of quantum chromodynamics? This is the million-dollar question, literally. The Clay Mathematics Institute has announced a million-dollar prize for a rigorous mathematical proof that quantum chromodynamics does not allow free quarks or gluons to be produced. While no claimants to the prize have yet come forward, we nevertheless have strong indirect support of this idea, coming not only from experimental observations, but also from numerical simulations that closely approximate the complicated interactions in quantum chromodynamics. This is heartening, if not definitive. We still have to confirm that it is some property of the theory and not of the computer simulation. However, for physicists, if not mathematicians, this seems pretty convincing.

One final bit of direct evidence that QCD is correct came from a realm where exact calculations *can* be done. Because quarks are not completely free at short distances, I earlier mentioned that there should be calculable corrections to exotic scaling phenomena observed in the high-energy collisions of electrons off protons and neutrons, as originally observed at SLAC. Perfect scaling would require completely noninteracting particles. The corrections that one could calculate in quantum chromodynamics would only be observable in experiments that were far more sensitive than those originally performed at SLAC. It took the development of new, higher-energy accelerators to probe them. After thirty years or so, enough evidence was in so that comparison of theoretical predictions and experiment agreed at the 1 percent level, and quantum chromodynamics as the theory of the strong interaction was finally verified in a precise and detailed way.

Gross, Wilczek, and Politzer were finally awarded the Nobel Prize in 2004 for their discovery of asymptotic freedom. The experimentalists who had first discovered scaling at SLAC, which was the key observation that set theorists off in the right direction, were awarded the Nobel Prize much earlier, in 1990. And the experimentalists who discovered the charmed quark in 1974 won the Nobel Prize two years later, in 1976.

But the biggest prize of all, as Richard Feynman has said, is *not* the recognition by a medal or a cash award, or even the praise one gets from colleagues or the public, but the prize of actually learning something new about nature.

. . .

In this sense the 1970s were perhaps the richest decade in the twentieth century, if not in the entire history of physics. In 1970 we understood only one force in nature completely as a quantum theory, namely quantum electrodynamics. By 1979 we had developed and experimentally verified perhaps the greatest theoretical edifice yet created by human minds, the Standard Model of particle physics, describing precisely three of the four known forces in nature. The effort spanned the entire history of modern science, from Galileo's investigations of the nature of moving bodies, through Newton's discovery of the laws of motion, through the experimental and theoretical investigations of the nature of electromagnetism, through Einstein's unification of space and time, through the discoveries of the nucleus, quantum mechanics, protons, neutrons, and the discovery of the weak and strong forces themselves.

But the most remarkable characteristic of all in this long march toward the light is how different the fundamental nature of reality is from the shadows of reality that we experience every day, and in particular how the fundamental quantities that appear to govern our existence are not fundamental at all.

Making up the heart of observed matter are particles that had never been directly observed and, if we are correct, will never be directly observable—quarks and gluons. The properties of forces that govern the interactions of these particles—and also the particles that have formed the basis of modern experimental physics for more than a century, electrons—are also, on a fundamental level, completely different from the properties we directly observe and on which we depend for our existence. The strong interaction between protons and neutrons is only a long-distance remnant of the underlying force between quarks, whose fundamental properties are masked by the complicated interactions within the nucleus. The weak interaction and the electromagnetic interaction, which could not be more different on the surface—one is short-range, while the other is long-range, and one appears thousands of times weaker than the other—are in fact intimately related and reflect different facets of a single whole.

That whole is hidden from us because of the accident of nature we call spontaneous symmetry breaking, which distinguishes the two weak and electromagnetic interactions in the world of our experience and hides their true nature. More than that, the properties of the particles that produce the characteristics of the beautiful world we observe around us are only possible because, after the accident of spontaneous symmetry breaking, just one particle in nature—the photon—remains massless. If symmetry breaking had never occurred so that underlying symmetries of the forces governing matter were manifest—which in turn would mean that the particles conveying the weak force would also be massless, as would most of the particles that make us up—essentially nothing we see in the universe today, from galaxies to stars, to planets, to people, to birds and bees, to scientists and politicians, would ever have formed.

Moreover, we have learned that even these particles that make us up are not all that exist in nature. The observed particles combine in simple groupings, or families. The up and down quarks make up protons and neutrons. Along with them one finds the electron, and its partner, the electron neutrino. Then, for reasons we still don't understand, there is a heavier family, made up of the charm and strange quark on the one hand, and the muon and its neutrino on the other. And finally, as experiments have now confirmed over the past decade or two, there is a third family, made of two new types of quarks, called bottom and top, and an accompanying heavy version of the electron called the tau particle, along with its neutrino.

Beyond these particles, as I shall soon describe, we have every reason to expect that other elementary particles exist that have never been observed. While these particles, which we think make up the mysterious dark matter that dominates the mass of our galaxy and all observed galaxies, may be invisible to our telescopes, our observations and theories nevertheless suggest that galaxies and stars could never have formed without the existence of dark matter.

And at the heart of all of the forces governing the dynamical behavior of everything we can observe is a beautiful mathematical framework called gauge symmetry. All of the known forces, strong, weak, electromagnetic, and even gravity, possess this mathematical property, and for the three former examples, it is precisely this property that ensures that the theories make mathematical sense and that nasty quantum infinities disappear from all calculations of quantities that can be compared to experiment.

With the exception of electromagnetism, these other symmetries remain completely hidden from view. The gauge symmetry of the strong force is hidden because confinement presumably hides the fundamental particles that manifest this symmetry. The gauge symmetry of the weak force is not manifest in the world in which we live because it is spontaneously broken so that the W and Z particles become extremely massive.

. . .

The shadows on the wall of everyday life are truly merely shadows. In this sense, the greatest story every told, so far, has been slowly playing out over the more than two thousand years since Plato first imagined it in his analogy of the cave.

But as remarkable as this story is, two elephants remain in the room. Two protagonists in our tale could until recently have meant that the key aspects of the story comprised a mere fairy tale invented by theorists with overactive imaginations.

First, the W and Z particles, postulated in 1960 to explain the weak interaction, almost one hundred times more massive than protons and neutrons, were still mere theoretical postulates, even if the indirect evidence for their existence was overwhelming. More than this, an invisible field—the Higgs field—was predicted to permeate all of space, masking the true nature of reality and making our existence possible because it spontaneously breaks the symmetry between the weak and the electromagnetic interactions.

To celebrate a story that claims to describe how it is that we exist, but that also posits an invisible field permeating all of space, sounds suspiciously like a religious celebration, and not a scientific one. To truly ensure that our beliefs conform to the evidence of reality rather than how we would *like* reality to be, to keep science worthy of the name, we *had* to discover the Higgs field. Only then could we truly know if the significance of the features of our world that we hold so dear might be no greater than that of the features of one random ice crystal on a window. Or, more to the point, perhaps, no greater than the significance of the superconducting nature of wire in a laboratory versus the normal resistance of the wires in my computer.

The experimental effort to carry out this task was no easier than that in developing the theory itself. In many ways it was more daunting, taking more than fifty years and involving the most difficult fabrication of technology that humans have ever attempted.

Chapter 20

SPANKING THE VACUUM

If anyone slaps you on the right cheek, turn to him the other also.

—MATTHEW 5:39

As the 1970s ended, theorists were on top of the world, triumphant and exultant. With progress leading to the Standard Model so swift, what other new worlds were there to conquer? Dreams of a theory of everything, long dormant, began to rise again and not just in the dim recesses of the collective subconscious of theorists.

Still, the W and Z gauge particles had never actually been observed, and the challenge to directly observe them was pretty daunting. Their masses were precisely predicted in the theory at about ninety times the mass of the proton. The challenge to produce these particles comes from a simple bit of physics.

Einstein's fundamental equation of relativity, $E = mc^2$, tells us that we can convert energy into mass by accelerating particles to energies of many times their rest mass. We can then smash them into targets to see what comes out.

The problem is that the energy available to produce new particles by smashing other fast-moving particles into stationary targets is given by what is called the center-of-mass energy. For those undaunted by an-other formula, this turns out to be the square root of twice the product of the energy of the accelerated particle times the rest mass energy of the target particle. Imagine accelerating a particle to one hundred times the rest mass energy of the proton (which is about one gigaelectronvolt—GeV). In a collision with stationary protons in a target, the center-of-mass energy that is available to create new particles is then only about 14 GeV. This is just slightly greater than the center-of-mass energy available in the highest-energy particle accelerator in 1972.

To reach the energies required to produce massive particles such as the W or Z bosons, two opposing beams of particles must collide. In this case the total center-of-mass energy is simply twice the energy of each beam. If each colliding beam of particles has an energy of one hun-dred times the rest mass of a proton, this then yields 200 GeV of energy to be converted into the mass of new particles.

Why, then, produce accelerators with stationary targets and not col-liders? The answer is quite simple. If I am shooting a bullet at a barn door, I am more or less guaranteed to hit something. If I shoot a bullet at another incoming bullet, however, I'd have to be a much better shot than probably anyone else alive and have a better gun than any now made to be guaranteed to hit it.

This was the challenge facing experimentalists in 1976, by which time they took the electroweak model seriously enough that they thought it worth the time, effort, and money to try to test it.

But no one knew how to build a device with the appropriate energy. Accelerating individual beams of particles or antiparticles to high ener-gies had been achieved. By 1976 protons were being accelerated to 500 GeV, and electrons up to 50 GeV. At lower energies, collisions of elec-trons and their antiparticles had successfully been carried out, and this is how the new particle containing the charmed quark and antiquark had been discovered in 1974.

Protons, having greater mass and thus more rest energy initially, are easier to accelerate to high energies. In 1976 a proton accelerator at the European Organization for Nuclear Research (CERN) in Geneva, the Super Proton Synchrotron (SPS), had just been commissioned as a conventional fixed-target accelerator operating with a proton beam at 400 GeV. However, another accelerator at Fermilab, near Chicago, had already achieved proton beams of 500 GeV by the time the SPS turned on. In June of that year, physicists Carlo Rubbia, Peter McIntyre, and David Cline made a bold suggestion at a neutrino conference: converting the SPS at CERN into a machine that collided protons with their antiparticles—antiprotons—would allow CERN to potentially produce W's and Z's.

Their bold idea was to use the same circular tunnel to accelerate protons in one direction, and antiprotons in another. Since the two particles have opposite electric charges, the same accelerating mechanism would have opposite effects on each particle. So a single accelerator could in principle produce two high-energy beams circulating in opposite directions.

The logic of such a proposal was clear, but its implementation was not. In the first place, given the strength of the weak interaction, the production of even a few W and Z particles would require the collision of hundreds of billions of protons and antiprotons. But no one had ever produced and collected enough antiprotons to make an accelerator beam.

Next, you might imagine that with two beams traversing the same tunnel in opposite directions, particles would be colliding all around the tunnel and not in the detectors designed to measure the products of the collisions. However, this is far from the case. The cross section of even a small tunnel compared to the size of the region over which a proton and an antiproton might collide is so huge that the problem is quite the opposite. It seemed impossible to produce enough antiprotons and ensure that both they and the protons in the proton beam would be sufficiently compressed so that when the two beams were brought together, steered by powerful magnets, any collisions at all would be observed.

Convincing the CERN directorate to transform one of the world's most powerful accelerators, built in a circular tunnel almost eight kilometers around at the French-Swiss border, into a new kind of collider would have been difficult for many people, but Carlo Rubbia, a bombastic force of nature, was up to the task. Few people who got in Rubbia's way were likely to be happy about it afterward. For eighteen years he jetted every week between CERN and Harvard, where he was a professor. His office was two floors down from mine, but I knew when he was in town because I could hear him. Moreover, Rubbia's idea was good, and in promoting it he was really suggesting to CERN that the SPS move up from an "also-ran" machine to the most exciting accelerator in the world. As Sheldon Glashow said to the CERN directorate when encouraging them to move forward, "Do you want to walk, or do you want to fly?"

Still, to fly one needs wings, and the creation of a new method to produce, store, accelerate, and focus a beam of antiprotons fell to a brilliant accelerator physicist at CERN, Simon van der Meer. His method was so clever that many physicists who first heard about it thought it violated some fundamental principles of thermodynamics. The properties of the particles in the beam would be measured at one place in the circular tunnel, then a signal would be sent for magnets farther down the tunnel to give many small kicks over time to the particles in the beam as they passed by, thus slightly altering the energies and momenta of any wayward particles so that they would eventually all get focused into a narrow beam. The method, called stochastic cooling, helped make sure particles that were wandering away from the center of the beam would be sent back into the middle.

Together van der Meer and Rubbia pushed forward, and by 1981 the collider was working as planned, and Rubbia assembled the largest physics collaboration ever created and built a large detector capable of sorting through billions of collisions of protons and antiprotons to search for a handful of possible W and Z particles. Rubbia's team was not the only one hunting for a W and a Z, however. Another detector collaboration had been assembled and was also built at CERN. Redundancy for such an important observation seemed appropriate.

Unearthing a signal from the immense background in these experiments was not easy. Remember that protons are made of more than one quark, and in a single proton-antiproton collision a lot of things can happen. Moreover, the W's and Z's would not be observed directly, but via their decays—in the case of the W, into electrons and neutrinos. Neutrinos would not be directly observed, either. Rather the experimentalists would tally up the total energy and momentum of each outgoing particle in a candidate event and look for large amounts of "missing energy," which would signal that a neutrino had been produced.

By December 1982, a W candidate event had been observed by Rubbia and his colleagues. Rubbia was eager to publish a paper based on this single event, but his colleagues were more cautious, for good reason. Rubbia seemed to have a history of making discoveries that weren't always there. In the meantime he leaked details of the event to a number of colleagues around the world.

Over the next few weeks his "UA1" collaboration obtained evidence for five more W candidate events, and the UA1 physicists designed several far more stringent tests to ascertain with high confidence that the candidates were real. On January 20, 1983, Rubbia presented a memorable and masterful seminar at CERN announcing the result. The standing ovation he received made it clear that the physics community was convinced. A few days later Rubbia submitted a paper to the journal *Physics Letters* announcing the discovery of six W events. The W had been discovered with precisely the predicted mass.

The search was not over, however. The Z remained to be seen. Its predicted mass was slightly higher than that of the W, and its signal was therefore slightly harder to obtain. Nevertheless, within a month or so of the W announcement, evidence for Z events began to come in from both experiments, and on the basis of a single clear event, on May 27 that year Rubbia announced its discovery.

The gauge bosons of the electroweak model had been found. The significance of these discoveries for solidifying the empirical basis of the Standard Model was underscored when, just slightly over a year after making the announcement, Rubbia and his accelerator colleague van der Meer were awarded the Nobel Prize in Physics. While the teams that had built and operated both the accelerator and the detectors were huge, few could deny that without Rubbia's drive and persistence and van der Meer's ingenious invention the discovery would not have been possible.

One big Holy Grail now remained: the purported Higgs particle. Unlike the W and Z bosons, the mass of the Higgs is *not* fixed by the theory. Its couplings to matter and to the gauge bosons were predicted, as these couplings allow the background Higgs field that presumably exists in nature to break the gauge symmetry and give mass not to just the W and the Z, but also to electrons, muons, and quarks—indeed to all the fundamental particles in the Standard Model save the neutrino and the photon. However, neither the Higgs particle mass nor the strength of its self-interactions was separately determined in advance by then existing measurements. Only their ratio was fixed by the theory in terms of the measured strength of the weak interaction between known particles.

Given conservative estimates of the possible magnitude of the Higgs self-interaction strength, the Higgs particle mass was conservatively estimated to lie within a range of 2 to 2,000 GeV. What set the upper limit was that, if the Higgs self-coupling is too big, then the theory becomes strongly interacting and many of the calculations performed using the simplest picture of the Higgs break down.

Aside from their necessary role in breaking the electroweak symmetry and giving other elementary particles masses, these quantitative details were therefore largely undetermined by experiments up to that time—which is probably why Sheldon Glashow in the 1980s referred to the Higgs as the "toilet" of modern physics. Everyone was aware of its necessary existence, but no one wanted to talk about the details in public.

That the Standard Model didn't fix in advance many of the details of the Higgs field didn't dissuade many theorists from proposing models that "predicted" the Higgs mass based on some new theoretical ideas. In the early 1980s, each time accelerators increased their energies, new physics papers would come out predicting a Higgs would be discovered when the machine was turned on. Then a new threshold would be reached, and nothing would be observed. To explore all the available parameter space to see if the Higgs existed, a radically new accelerator would clearly have to be built.

I was convinced during all this time that the Higgs didn't exist. The spontaneous symmetry breaking of the electroweak gauge symmetry did certainly occur—the W and the Z exist and have mass—but adding a fundamental new scalar field designed by recipe specifically to perform this task seemed contrived to me. First, no other fundamental scalar field had ever been observed to exist in nature's particle menagerie. Second, I felt that with all of the unknown physics yet to be discovered at small scales, nature would have developed a much more ingenious and unexpected way of breaking the gauge symmetry. Once one posits the Higgs particle, then the next obvious question is "Why that?" or more specifically "Why just the right dynamics to cause it to condense at that scale, and with that mass?" I thought that nature would find a way to break the theory in a less ad hoc fashion, and I expressed this conviction fairly strongly when I was interviewed for my eventual position at the Society of Fellows at Harvard after getting my PhD.

Let's recall now what the existence of the Higgs implies. It requires not just a new particle in nature but an invisible background field that must exist throughout all of space. It also implies that all particles—not just the W and the Z particles but also electrons and quarks—are massless in the fundamental theory. These particles that interact with the Higgs background field then experience a kind of resistance to their motion that slows their travel to less than the speed of light—just as a swimmer in molasses will move more slowly than a swimmer in water.

Once they are moving at sub-light-speed, the particles behave as if they are massive. Those particles that interact more strongly with this background field will experience a greater resistance and will act as if they are more massive, just as a car that goes off the road into mud will be harder to push than if it were on the pavement, and to those pushing it, it will seem heavier.

This is a remarkable claim about the nature of reality. Remembering that in superconductors the condensate that forms is a complicated state of bound pairs of electrons, I was skeptical that things would work out so much more simply and cleanly on fundamental scales in empty space.

So how to explore such a remarkable claim? We use the central property of quantum field theory that was exploited by Higgs himself when he proposed his idea. For every new field in nature, at least one new type of elementary particle must exist with that field. How, then, to produce the particles if such a background field exists throughout space?

Simple. We *spank* the vacuum.

By this I mean that if we can focus enough energy at a single point in space, we can excite real Higgs particles to emerge and be measured. One can picture this as follows. In the language of elementary particle physics, using Feynman diagrams, we can think of a virtual Higgs particle emerging from the background Higgs field, giving mass to other particles. The left diagram corresponds to particles such as quarks and electrons scattering off a virtual Higgs particle and being deflected, thus experiencing resistance to their forward motion. The right diagram represents the same effect for particles such as the W and the Z.

We can then simply turn this picture around:

In this case energetic particles such as W's and Z's or quarks and/or antiquarks or electrons and/or positrons appear to emit virtual Higgs particles and recoil. If the energies of the incoming particles are large enough, then the emitted Higgs could be a real particle. If they aren't, the Higgs would be a virtual particle.

Now remember that if the Higgs gives mass to particles, then the particles it interacts with most strongly will be the particles that get the largest masses. In turn this means that the particles most likely to spit out a Higgs are the incident particles with the heaviest masses. That means that light particles such as electrons are probably not a good bet to directly create Higgs particles in an accelerator. Instead we can imagine creating an accelerator with enough energy so that we can create heavy virtual particles that will spit out Higgs particles, either virtual or real.

The natural candidates are then protons. Build an accelerator or a collider starting with protons and accelerate them to high enough energies to produce enough virtual heavy constituents so as to produce Higgs particles. The Higgs particles, virtual or real, being heavy, will quickly decay into the lighter particles that the Higgs interacts with most strongly—once again either the top or bottom quarks or W's and Z's. These will in turn decay into other particles.

The trick would be to consider events with the smallest number of outgoing particles that could be cleanly detected, then determine their energies and momenta precisely and see if one could reconstruct a series of events traceable to a single massive intermediate particle with the predicted interactions of a Higgs particle. No small task!

These ideas were already clear as early as 1977, even before the discovery of the top quark itself (since the bottom quark had already been discovered, and all the other quarks came in weak pairs—up and down, charm and strange—clearly another quark had to exist, although it took until 1995 to discover it, a whopping 175 times heavier than the proton). But knowing what was required and actually building a machine capable of doing the job were two different things.

Chapter 21

GOTHIC CATHEDRALS OF THE TWENTY-FIRST CENTURY

The price of wisdom is above rubies.

—JOB 28:18

$\mathbb{A}$ccelerating protons to high enough energies to explore the full range of possible Higgs masses was well beyond the capabilities of any machine in 1978—when all the other predictions of the electroweak theory were confirmed—or in 1983 when the W and the Z had been discovered. An accelerator at least an order of magnitude more powerful than the most powerful machine then in existence was required. In short, not a collider, but a supercollider.

The United States, which for the entire period since the end of the Second World War had dominated science and technology, had good reason to want to build such a machine. After all, CERN in Geneva had emerged by 1984 as the dominant particle physics laboratory in the world. American pride was so hurt when both the W and the Z particles were discovered at CERN that six days after the press conference announcing the Z discovery, the *New York Times* published an editorial titled "Europe 3, U.S. Not Even Z-Zero"!

Within a week after the Z discovery, American physicists decided to cancel construction of an intermediate-scale accelerator in Long Island and go for broke. They would build a massive accelerator with a center-of-mass energy almost one hundred times greater than the CERN SPS machine. To do so they would need new superconducting magnets, and so their brainchild was named the Superconducting Super Collider (SSC).

After the project was proposed by the US particle physics community in 1983, the traditional scramble proceeded among many different states to get a piece of the enormous fiscal pie for its construction and management. After much political and scientific wrangling a site just south of Dallas, Texas, in Waxahachie, was chosen. Whatever the motivation, Texas seemed particularly appropriate, as everything about the project, which was approved in 1987 by President Reagan, was supersize.

The huge underground tunnel would have been eighty-seven kilometers around, the largest tunnel ever constructed. The project would be twenty times bigger than any other physics project ever attempted. The proposed energy of collisions, with two beams each having an energy twenty thousand times the mass of the proton, would be about one hundred times larger than the collision energy of the machine at CERN that had discovered the W and the Z. Ten thousand superconducting magnets, each of unprecedented strength, would have been required.

Cost overruns, lack of international cooperation, a poor US economy, and political machinations eventually led to SSC's demise in October 1993. I remember the time well. I had recently moved from Yale to become chair of the Physics Department at Case Western Reserve University, with a mandate to rebuild the department and hire twelve new faculty members over five years. The first year we advertised, in 1993–94, we received more than two hundred applications from senior scientists who had been employed at the SSC and who were now without a job or any prospects. Many of them were very senior, having left full professorships at distinguished universities to spearhead the effort. It was sad, and more than half of those people had to leave the field altogether.

The anticipated cost of the project when it was canceled had risen from $4.4 billion at its inception in 1987 to about $12 billion in 1993. While this was, and still is, a large amount of money, one can debate the merits of killing the project. Two billion dollars had already been spent on it, and twenty-four kilometers of tunnel had been completed.

The decision to kill the project was not black-and-white, but a number of things could have played a bigger role in considerations—from the opportunity costs of losing a fair fraction of the talented accelerator physicists and particle physics experimentalists in the country to the many new breakthroughs that might have resulted from the expenditures on high-tech development that would have contributed to our economy. Moreover, had the SSC been built and functioned as planned, we may have had answers more than a decade ago to experimental questions we are still addressing. Would knowing the answers have changed anything we might have done in the meantime? We'll probably never know.

The $12 billion would have been spent over some ten to fifteen years during construction and the commencement of operations, which makes the cost in the range of $1 billion per year. In the federal budget this is not a large amount. My own political views are well known, so it may not be surprising for me to suggest, for example, that the United States would have been just as secure had it cut the bloated US defense budget by this amount, far less than 1 percent of its total each year. Moreover, the entire cost of the SSC would have probably been comparable to the air-conditioning and transportation costs of the disastrous 2003 Iraq invasion, which decreased our net security and well-being. I can't help referring once again to Robert Wilson's testimony before Congress regarding the Fermilab accelerator: "It has nothing to do directly with defending our country except to help make it worth defending."

These are political questions, however, not scientific ones, and in a democracy, Congress, representing the public, has the right and responsibility to oversee priorities for expenditures on large public projects.

The particle physics community, perhaps too used to a secure inflow of money during the Cold War, did not do an adequate job of informing the public and Congress what the project was all about. It is not surprising that in hard economic times the first thing to be cut would be something that seemed so esoteric. I wondered at the time why it was necessary to kill the project, rather than suspend funding until the economy improved or until technological developments might have reduced its cost. Neither the tunnel (now filling with water) nor the laboratory buildings (now occupied by a chemical company) were going anywhere.

Despite these developments in the United States, CERN was moving forward with a new machine, the Large Electron-Positron (LEP) Collider, designed to explore in detail the physics of the W and the Z particles, at the urging of its newest Nobel laureate, the indomitable Carlo Rubbia. He became the laboratory's director in 1989, the same year the new machine came online.

A twenty-seven-kilometer-long circular tunnel was dug about a hundred meters underground around the old SPS machine, which was now used to inject electrons and positrons into the bigger ring, where they were further accelerated to huge energies. Located on the outskirts of Geneva, the new machine was large enough to cross under the Jura Mountains into France. European nations are more familiar with building tunnels than the United States is, and when the tunnel was completed, the two ends met up to within one centimeter. Moreover CERN, as an international collaboration of many countries, did not significantly eat into the GDP of any one country.

The new machine ran successfully for more than a decade, and after the demise of the SSC in the United States, the huge LEP tunnel was considered for the creation of a miniversion of the SSC—not quite as powerful but still energetic enough to explore much of the parameter space where the long-sought Higgs particle might exist. Some competition came from a machine at Fermilab, called the Tevatron, which had been running since 1976 and in 1984 came online as the world's most energetic proton-antiproton machine. By 1986, the collision energy of protons and antiprotons circulating around the 6.5 kilometer ring of superconducting magnets at Fermilab was almost two thousand times the equivalent rest mass energy of the proton.

As significant as this was, it was not sufficient to probe most of the available parameter space for the Higgs, and a discovery at the Tevatron would have required nature to have been kind. The Tevatron did garner one great success, the long-anticipated discovery, in 1995, of the mammoth top quark, 175 times the mass of the proton, and the most massive particle yet discovered in nature.

With no clear competition therefore, within fourteen months of the demise of the SSC the CERN council approved the construction of a new machine, the Large Hadron Collider, in the LEP tunnel. Design and development of the machine and detectors would take some time to complete, so the LEP machine would continue to operate in the tunnel for almost another six years before having to close down for reconstruction. It would then take almost another decade to complete construction of the machine and the particle detectors to be used in the search for the Higgs and/or other new physics.

That is, *if* a working machine and viable detectors could be constructed. This would be the most complicated engineering task humans had ever undertaken. The design specifications for superconducting magnets, computing facilities, and many other aspects of the machine and detectors called for technology far exceeding anything then available.

Conceptual design of the machine took a full year, and another year later two of the main experimental detector collaboration proposals were approved. The United States, with no horses in this race, was admitted as an "observer" state to CERN, allowing US physicists to become key players in detector development and design. In 1998 construction of the cavern to hold one of the two major detectors, the CMS detector, was delayed for six months as workers discovered fourth-century Gallo-Roman ruins, including a villa and surrounding fields, on the site.

Four and a half years later, the huge caverns that would house both main detectors underground were completed. Over the next two years, 1,232 huge magnets, each fifteen meters long and weighing thirty-five tons, were lowered fifty meters below the surface in a special shaft and delivered to their final destinations using a specially designed vehicle that could travel in the tunnel. A year after that, the final pieces of each of the two large detectors were lowered into place, and at 10:28 a.m., September 10, 2008, the machine officially turned on for the first time.

Two weeks later, disaster struck. A short occurred in one of the magnet connectors, causing the associated superconducting magnet to go normal, releasing a huge amount of energy and resulting in mechanical damage and release of some of the liquid helium cooling the machine. The damage was extensive enough that a redesign and examination of every weld and connection in the LHC was required, taking more than a year to complete. In November of 2009 the LHC was finally turned back on, but because of design concerns, it was set to run at seven thousand times the center-of-mass energy of the proton, instead of fourteen thousand. On March 19, 2010, the machine finally began running with colliding beams at the lower energy, and both sets of detectors began to record collisions with this total energy within two weeks.

These simple timelines belie the incredible challenges of the technical feats achieved at CERN during the fifteen years since the machine was first proposed. If you land at Geneva airport and look outside, you will see gentle farmland, with mountains in the distance. Without being told, no one would guess that underneath that farmland lies the most complicated machine humans have ever constructed. Consider some of the characteristics of the machine, which lies at some points 175 meters below this calm and pastoral scene:

1. In the 3.8-meter-wide tunnel, traversing twenty-seven kilometers, are two parallel beamline circles, intersecting at four points around the ring. Distributed around the ring are more than sixteen hundred superconducting magnets, most weighing more than twenty-seven tons. The tunnel is so long that, looking down it, one almost cannot see its curvature:

2. Ninety-six tons of superfluid ^{4}He are used to keep the magnets operating at a temperature of less than two degrees above absolute zero, colder than the temperature of the radiation background in the depths of interstellar space. In total, 120 tons of liquid helium are utilized, cooled first by using about ten thousand tons of liquid nitrogen. Some forty thousand leak-tight pipe connections had to be made. The volume of He used makes the LHC the largest cryogenic facility in the world.

3. The vacuum in the beamlines is required to be sparser than the vacuum in outer space experienced by the astronauts performing space walks outside the ISS, and ten times lower than the atmospheric pressure on the Moon. The largest volume at the LHC pumped down to this vacuum level is nine thousand cubic meters, comparable to the volume of a large cathedral.

4. The protons accelerated around the tunnel in either direction move at a speed of 0.999999991 times the speed of light, or only

about three meters per second less than light speed. The energy possessed by each proton in the collision is equivalent to the energy of a flying mosquito, but compressed into a radial dimension one million million times smaller than a mosquito's length.

5. Each beam of protons is bunched into 2,808 separate bunches, squeezed at collision points to about one-quarter the width of a human hair, around the ring, with 115 billion protons in each bunch, yielding bunch collisions every twenty-five-billionths of a second, with more than 600 million particle collisions per second.

6. The computer grid designed to handle data from the LHC is the largest in the world. Every second the raw data generated by the LHC are enough to fill more than a thousand one-terabyte hard drives. This must be reduced considerably to be analyzed. From the 6 million billion proton-proton collisions analyzed in 2012 alone, more than twenty-five thousand terabytes of data were processed—more than the amount of information in all the books ever written and corresponding to a stack of CDs about twenty kilometers tall. To do this, a worldwide computer grid was created with 170 computer centers in thirty-six countries. When the machine is running, about seven hundred megabytes per second of data are produced.

7. The requirements for the sixteen hundred magnets to produce beams intense enough to collide is equivalent to firing two needles from a distance of ten kilometers with such precision that they collide exactly halfway between the two firing positions.

8. The alignment of the beams is so precise that account must be taken for the tidal variations on the ring from the gravity of the Moon as its position over Geneva changes, causing a variation of one millimeter in the circumference of the LHC each day.

9. To produce the incredibly intense magnetic fields needed to steer the proton beams, a current of almost twelve thousand amps flows through each of the superconducting magnets, about 120 times the current flowing through an average family house.

10. The strands of cable needed to make up the magnetic coils in the LHC span about 270,000 kilometers, or about six times the circumference of the Earth. If all the filaments in the strands were unraveled, they would stretch to the Sun and back more than five times.

11. The total energy in each beam is about the same as that of a four-hundred-ton train traveling at 150 km/hr. This is enough energy to melt five hundred kilograms of copper. The energy stored in the superconducting magnets is thirty times higher than this.

12. Even with the superconducting magnets—which make power consumption in the machine manageable—when the machine is running, it uses about the same power as the total consumption of all of the households in Geneva.

So much for the machine itself. To analyze the collisions at the LHC, a variety of large detectors have been built. Each of the four currently operating detectors has the size of a significant office building and the complexity of a major laboratory. To have the opportunity to go underground and see the detectors is to feel like Gulliver in Brobdingnag. The scale of absolutely every component is immense. Here is a photo of the CMS detector, the smaller of the two largest detectors at the LHC:

If you are actually at the detector, it is hard to even grasp the full picture, as can be seen in the more up-close-and-personal view:

The complexity of the machines is almost unfathomable. For a theorist such as me, it is hard to imagine how any single group of physicists can keep track of the device, much less design and build it to the exacting specifications required.

Each of the two largest detectors, ATLAS and CMS, was built by a collaboration of over two thousand scientists. More than ten thousand scientists and engineers from over a hundred countries participated in building the machine and detectors. Consider the smaller of the two detectors, CMS. It is more than twenty meters long, fifteen meters high, and fifteen meters wide. Some 12,500 tons of iron are in the detector, more than in the Eiffel Tower. The two halves of the detector are separated by a few meters when it is being worked on. Even though they are not on wheels, if the two halves were apart when the large magnetic field of the detector was turned on, they would be dragged together.

Each detector is separated into millions of components, with trackers that can measure particle trajectories to an accuracy of ten-millionths of a meter, with calorimeters, which detect to a high accuracy energy deposited in the detectors, and with devices for measuring the speed of particles by measuring the radiation they emit as they traverse the detector. In each collision hundreds or thousands of individual particles may be produced, and the detector must keep track of almost all of them to reconstruct each event.

Physicist Victor Weisskopf was the fourth director general of CERN, between 1961 and 1966, and he likened the great accelerators of that time to the Gothic cathedrals of medieval Europe. In thinking of CERN and the LHC, the comparison is particularly interesting.

The Gothic cathedrals stretched the technology of the time, requiring new building techniques and new tools to be created. Hundreds or thousands of master craftsmen from dozens of countries built them over many decades. Their scale dwarfed that of any buildings that had previously been created. And they were built for no more practical reason than to celebrate the glory of God.

The LHC is the most complicated machine ever built, requiring new building techniques and new tools to be created. Thousands of PhD scientists and engineers from hundreds of countries speaking dozens of languages, and hailing from a background of at least an equal number of religions, were required to build the accelerator and the detectors that monitor it—taking almost two decades to complete the task. Its scale dwarfs that of all machines constructed before it. And it was built for no more practical reason than to celebrate and explore the beauty of nature.

Seen in this perspective, the cathedrals and the collider are both monuments to what may be best about human civilization—the ability and the will to imagine and construct objects of a scale and complexity that requires the cooperation of countless individuals, from around the globe if necessary, for the purpose of turning our awe and wonder at the workings of the cosmos into something concrete that may improve the human condition. Colliders and cathedrals are both works of incomparable grandeur that celebrate the human experience in different realms. Nevertheless, I think the LHC wins, and its successful construction over two decades demonstrates that the twenty-first century is not yet devoid of culture and imagination.

Which brings me finally to the road to July 4, 2012.

By 2011 the LHC was cruising along, as one of the CERN officials put it. The amount of data taken by October of that year was already 4 million times higher than during the first run in 2010, and thirty times higher than had been obtained by the beginning of 2011.

At this point in the collection of data that physicists had been waiting forty years for, rumors began to fly in the community. Many of these came from the experimenters themselves. I have a part-time position at Australian National University in Canberra, and the International Conference on High Energy Physics was going to be in held in Melbourne in July of 2012. Melbourne has a big LHC contingent, and when visiting, I kept hearing how a greater and greater possible mass range for the Higgs particle had been ruled out by the experiments already.

Many experimentalists relish being able to prove theorists wrong. So it was in this case. One experimentalist had excitedly told me less than six months before the meeting that the entire Higgs mass region had been ruled out except for a narrow range between 120 and 130 times the mass of the proton. She expected that by July they would be able to rule out that region too. As one who was skeptical of the Higgs, I wasn't unhappy to hear this. In fact, I was getting a paper ready to explain why the Higgs might not exist.

On April 5, the situation got more interesting as the LHC center-of-mass beam energy was increased slightly, to eight thousand times the rest energy of the proton. This translated into an increased potential for new particle discovery. By mid-June it was announced that the leaders of the two main experiments, along with the director general of CERN, would not be traveling to Melbourne for the meeting, but would be presenting results remotely from a televised conference on the morning of July 4 in the main colloquium room at CERN—the same room where Rubbia had announced the discovery of the W particles.

On July 4 I was at a physics meeting in Aspen, Colorado. Because of the significance of the impending announcement, the physics community there had set up a live remote presentation screen—so that at 1:00 a.m. we could all sit and watch history unfold. About fifteen of us showed up in the dark at the Aspen Center for Physics, mostly physicists, but also a few journalists, including Dennis Overbye from the *New York Times*, who knew he was going to have a late night writing. As it turned out, so would I. The *Times* had asked me for an essay for the following week's Science Times section if things worked out as expected.

Then the show began, and in the next forty-five minutes or so spokespeople presented data from both of the two large detectors that compellingly demonstrated the existence of a new elementary particle with mass of about 125 times the mass of the proton. After the initial catastrophe in 2009, the LHC had functioned impeccably—as had both the detectors. I and many of my colleagues were amazed during the early months by the immaculately clean results the detectors displayed regarding known background processes. So we were not surprised that when something new appeared, these detectors could find it, in spite of the unbelievably complicated environment that the detectors were functioning in.

But more than this, the particle was discovered by looking precisely at the decay channels that had been predicted for a Standard Model Higgs particle. The relative decays into photons (via intermediate top quarks or W's) versus particles such as electrons (via intermediate Z bosons) agreed more or less with what was predicted, as did the production rate of the new particle in the proton-proton collisions. Of the billions and billions of collisions analyzed by the two detector collaborations up to that point, about fifty potential Higgs candidates had been discovered. Many tests needed to be performed to get a more definitive identification, but if it walked like a Higgs and quacked like a Higgs, it probably was a Higgs. The evidence was good enough that François Englert and Peter Higgs were awarded the Nobel Prize in October of 2013, the first year possible after the claimed discovery.

In February 2013, the LHC shut down and the machine was upgraded so that it could finally run at its originally designed energy and luminosity. By the final weeks before turnoff, the CERN mass-storage systems had stored more than one hundred petabytes of data, more info than in 100 million CDs. New results continued to roll in from data that had not yet been analyzed before the first announcement (including tantalizing hints of a possible new and unexpected heavy particle, six times heavier than the Higgs, hints that disappeared just as this book was being sent off to press).

For a real discovery, the more data you have, the better it looks, whereas anomalous results tend to disappear over time. This time things looked good, almost embarrassingly so. If one compared five different predicted decay channels into photons, Z particles, W particles, tau particles (the heaviest known cousin of the electron), and particles containing b quarks, to observation, the predictions of the Standard Model Higgs, with no extra accessories, agreed strikingly well.

From the angular distribution and energies of the decay products, with a new larger sample of Higgs candidates, the LHC detectors were able to explore whether the particle was indeed a scalar particle, which would make it the first fundamental scalar ever observed in nature. On March 26, 2015, the ATLAS detector at CERN released results that showed with greater than 99 percent confidence that the new particle was a spin 0 particle, with precisely the proper parity assignment to be a Higgs scalar. Nature had shown that it does not abhor scalar fields like the Higgs, as I for one had thought. The existence of such a fundamental scalar changes a great deal about what may be possible in nature, and people, including me, began to consider scenarios we would never before have considered.

In September 2015, about a month before the first draft of this book was written, the two large detectors ATLAS and CMS combined their data from 2011 and 2012 and presented for the first time a unified comparison of theory and experiment. The result—involving a mammoth computational effort to take into account separate systematic effects in each experiment, involving a total of forty-two hundred parameters—showed with a residual uncertainty of about 10 percent that the new particle had all the properties predicted for the Standard Model Higgs.

This simple conclusion may seem almost anticlimactic, following as it does a half century of directed effort by thousands of individuals—the theorists who developed the Standard Model and the others who performed the incredibly complex calculations needed to compare predictions with experiments, to determine background rates, and so on, and the thousands of experimental physicists who had built and tested and operated the most complex machine ever constructed. Their story was marked by incredible heights of intellectual bravery, years of confusion, bad luck and serendipity, rivalries and passion, and above all the persistence of a community focused on a single goal—to understand nature at her most fundamental scales. Like any human drama, it also included its share of envy, stubbornness, and vanity, but more important, it involved a unique community built completely independent of ethnicity, language, religion, or gender. It is a story that carries with it all the drama of the best epic tales and reflects the best of what science can offer to modern civilization.

That nature would be so kind as to actually use the ideas that a small collection of individuals wrote down on paper, inspired by abstract ideas of symmetry and using the complex mathematics of quantum field theory, will always seem to me nothing short of remarkable. It is hard to express the mixture of exhilaration and terror that comes from the realization that nature might actually work the way you are proposing it does when putting the final touches on a paper, possibly late at night, alone in your study. I suppose it may resemble the reaction Plato described that his poor philosophers might have as they are dragged out into the sunlight away from the cave for the first time.

To have discovered that nature really follows the simple and elegant rules intuited by the twentieth- and twenty-first-century versions of Plato's philosophers is both shocking and reassuring. It hints that the willingness of scientists to build an intellectual house of cards that could come tumbling down at the slightest experimental tremor was not misplaced. It gives us courage to continue to suppose, as Einstein had once expressed his amazement about, that the universe on its grandest scale is fathomable after all.

After witnessing the announcement of the Higgs discovery on July 4, 2012, I wrote the following:

The apparent discovery of the Higgs may not result in a better toaster or a faster car. But it provides a remarkable celebration of the human mind's capacity to uncover nature's secrets, and of the technology we have built to control them. Hidden in what seems like empty space—indeed, like nothing, which is getting more interesting all the time—are the very elements that allow for our existence.

By demonstrating this, last week's discovery will change our view of ourselves and our place in the universe. Surely that is the hallmark of great music, great literature, great art . . . and great science.

It is too early yet to judge or even fully anticipate what changes in our picture of reality will result from the Higgs discovery at the LHC, or the discoveries that may follow. Yet fortune does favor the prepared mind, and it is at once the responsibility and the joy of theorists such as me to ponder just that.

While nature may have appeared to be kind to us this time, perhaps it was too kind. The epic saga I have described here may yet provide a dramatic new challenge for physics and for physicists, and an explicit reminder that nature doesn't exist to make us comfortable. Because while we may have found what we expected, no one really expected to find just that and nothing else. . . .

Chapter 22

MORE QUESTIONS THAN ANSWERS

*A fool takes no pleasure in understanding, but only
in expressing his opinion.*

—PROVERBS 18:2

In one sense, our story might end here, because we have
come to the limits of our direct empirical knowledge about the universe
at its fundamental scales. But no one says we have to stop dreaming,
even if the dreams are not always pleasant. Before July 2012 particle
physicists had two nightmares. The first was that the LHC would see
precisely nothing. For if it did, it would likely be the last large accelerator
ever built to probe the fundamental makeup of the cosmos. The second
was that the LHC would discover the Higgs . . . period.

Each time we peel back one layer of reality, other layers beckon. So
each important new development in science generally leaves us with more
questions than answers. But it also usually leaves us with at least the out-
line of a road map to help us begin to seek answers to those questions. The
discovery of the Higgs particle, and with it the validation of the existence
of an invisible background Higgs field throughout space, was a profound
validation of the bold scientific developments of the twentieth century.

However, the words of Sheldon Glashow continue to ring true: The Higgs is like a toilet. It hides all the messy details we would rather not speak of. The Higgs field, as elegant as it might be, is within the Standard Model essentially an ad hoc addition. It is added to the theory to do what is required to accurately model the world of our experience. But it is *not* required by the theory. The universe could have happily existed with a long-range weak force and massless particles. We would just not be here to ask about them. Moreover, the detailed physics of the Higgs is, as we have seen, undetermined within the Standard Model alone. The Higgs could have been twenty times heavier, or a hundred times lighter.

Why, then, does the Higgs exist at all? And why does it have the mass it does? (Recognizing once again that whenever scientists ask "Why?," we really mean "How?") If the Higgs did not exist, the world we see would not exist, but surely that is not an explanation. Or is it? Ultimately to understand the underlying physics behind the Higgs is to understand how we came to exist. When we ask, "Why are we here?," at a fundamental level we may as well be asking, "Why is the Higgs here?" And the Standard Model gives no answer to this question.

Some hints do exist, however, coming from a combination of theory and experiment. Shortly after the fundamental structure of the Standard Model became firmly established, in 1974, and well before the details were experimentally verified over the next decade, two different groups of physicists at Harvard, where both Glashow and Weinberg were working, noticed something interesting. Glashow, along with Howard Georgi, did what Glashow did best: they looked for patterns among the existing particles and forces and sought out new possibilities using the mathematics of group theory.

Remember that in the Standard Model the weak and electromagnetic forces are unified at a high-energy scale, but when the symmetry is spontaneously broken by the Higgs field condensate, this leaves, at observable scales, two separate and distinct forces—with the weak force being short-range and electromagnetism remaining long-range. Georgi and

Glashow tried to extend this idea to include the strong force and discovered that all of the known particles and the three nongravitational forces could naturally fit within a single fundamental larger-gauge symmetry structure. They then speculated that this fundamental symmetry could spontaneously break at some ultrahigh energy and short-distance scale far beyond the range of current experiments, leaving two separate and distinct unbroken gauge symmetries left over—resulting in the separate strong and electroweak forces. Subsequently, at a lower energy and larger distance scale, the electroweak symmetry would break, separating that into the short-range weak and the long-range electromagnetic force.

They called such a theory, modestly, a Grand Unified Theory (GUT).

At around the same time, Weinberg and Georgi along with Helen Quinn noticed something interesting—following the work of Wilczek, Gross, and Politzer. While the strong interaction got weaker as one probed it at smaller-distance scales, the electromagnetic and weak interactions got stronger.

It didn't take a rocket scientist to wonder whether the strength of the three different interactions might become identical at some small-distance scale. When they did the calculations, they found (with the accuracy with which the interactions were then measured) that such a unification looked possible, but only if the scale of unification was about fifteen orders of magnitude in scale smaller than the size of the proton.

This was good news if the unified theory was the one proposed by Georgi and Glashow—because if all the particles we observe in nature got unified in this new large-gauge group, then new gauge bosons would exist that produce transitions between quarks (which make up protons and neutrons), and electrons and neutrinos. That would mean protons could decay into other lighter particles. As Glashow put it, "Diamonds aren't forever."

Even then it was known that protons must have an incredibly long lifetime. Not just because we still exist almost 14 billion years after the Big Bang, but because we all don't die of cancer as children. If protons decayed with an average lifetime smaller than about a billion billion years, then enough protons would still decay in our bodies during our childhood to produce enough radiation to kill us. Remember that in quantum mechanics, processes are probabilistic. If an average proton lives a billion billion years, then if one has a billion billion protons, on average one will decay each year. A lot more than a billion billion protons are in our bodies.

However, with the incredibly small proposed distance scale and therefore the incredibly large mass scale associated with spontaneous symmetry breaking in Grand Unification, the new gauge bosons would get large masses. That would make the interactions they mediate be so short-range that they would be unbelievably weak on the scale of protons and neutrons today. As a result, while protons could decay, they might live, in this scenario, perhaps a million billion billion billion years before decaying. No problem.

．．．

With the results of Glashow and Georgi, and Georgi, Quinn, and Weinberg, the smell of grand synthesis was in the air. After the success of the electroweak theory, particle physicists were feeling ambitious and ready for further unification.

How would one know if these ideas were correct, however? There was no way to build an accelerator to probe an energy scale a million billion times greater than the rest mass energy of protons. Such a machine would have to have a circumference of the Moon's orbit. Even if it was possible, considering the earlier debacle over the SSC, no government would ever foot the bill.

Happily, there was another way, using the kind of probability arguments I just presented that give limits to the proton lifetime. If the new Grand Unified Theory predicted a proton lifetime of, say, a thousand billion billion billion years, then if one could put a thousand billion billion billion protons in a single detector, on average one of them would decay each year.

Where could one find so many protons? Simple: in about three thousand tons of water.

So all that was required was to get a tank of, say, three thousand tons of water, put it in the dark, make sure there were no radioactivity backgrounds, surround it with sensitive phototubes that can detect flashes of light in the detector, and then wait for a year to see a burst of light when a proton decayed. As daunting as this may seem, at least two large experiments were commissioned and built to do just this, one deep underground next to Lake Erie in a salt mine, and one in a mine near Kamioka, Japan. The mines were necessary to screen out incoming cosmic rays that would otherwise produce a background that would swamp any proton decay signal.

Both experiments began taking data around 1982–83. Grand Unification seemed so compelling that the physics community was confident a signal would soon appear and Grand Unification would mean the culmination of a decade of amazing change and discovery in particle physics—not to mention another Nobel Prize for Glashow and maybe some others.

Unfortunately, nature was not so kind in this instance. No signals were seen in the first year, the second, or the third. The simplest elegant model proposed by Glashow and Georgi was soon ruled out. But once the Grand Unification bug had caught on, it was not easy to let it go. Other proposals were made for unified theories that might cause proton decay to be suppressed beyond the limits of the ongoing experiments.

On February 23, 1987, however, another event occurred that demonstrates a maxim I have found is almost universal: every time we open a new window on the universe, we are surprised. On that day a group of astronomers observed, in photographic plates obtained during the night, the closest exploding star (a supernova) seen in almost four hundred years. The star, about 160,000 light-years away, was in the Large Magellanic Cloud—a small satellite galaxy of the Milky Way observable in the southern hemisphere.

If our ideas about exploding stars are correct, most of the energy released should be in the form of neutrinos, despite that the visible light released is so great that supernovas are the brightest cosmic fireworks in the sky when they explode (at a rate of about one explosion per hundred years per galaxy). Rough estimates then suggested that the huge IMB (Irvine-Michigan-Brookhaven) and Kamiokande water detectors should see about twenty neutrino events. When the IMB and Kamiokande experimentalists went back and reviewed their data for that day, lo and behold IMB displayed eight candidate events in a ten-second interval, and Kamiokande displayed eleven such events. In the world of neutrino physics, this was a flood of data. The field of neutrino astrophysics had suddenly reached maturity. These nineteen events produced perhaps nineteen hundred papers by physicists, such as me, who realized that they provided an unprecedented window into the core of an exploding star, and a laboratory not just for astrophysics but also for the physics of neutrinos themselves.

Spurred on by the realization that large proton-decay detectors might serve a dual purpose as new astrophysical neutrino detectors, several groups began to build a new generation of such dual-purpose detectors. The largest one in the world was again built in the Kamioka mine and was called Super-Kamiokande, and with good reason. This mammoth fifty-thousand-ton tank of water, surrounded by 11,800 phototubes, was operated in a working mine, yet the experiment was maintained with the purity of a laboratory clean room. This was absolutely necessary because in a detector of this size one had to worry not only about external cosmic rays, but also about internal radioactive contaminants in the water that could swamp any signals being searched for.

Meanwhile, interest in a related astrophysical neutrino signature also reached a new high during this period. The Sun produces neutrinos due to the nuclear reactions in its core that power it, and over twenty years, using a huge underground detector, Ray Davis had detected solar neutrinos, but had consistently found an event rate about a factor of three below what was predicted using the best models of the Sun. A new type of solar neutrino detector was built inside a deep mine in Sudbury, Canada, which became known as the Sudbury Neutrino Observatory (SNO).

Super-Kamiokande has now been operating almost continuously, through various upgrades, for more than twenty years. No proton-decay signals have been seen, and no new supernovas observed. However, the precision observations of neutrinos at this huge detector, combined with complementary observations at SNO, definitely established that the solar neutrino deficit observed by Ray Davis is real, and moreover that it is not due to astrophysical effects in the Sun but rather due to the properties of neutrinos. At least one of the three known types of neutrinos is not massless—although it has a small mass indeed, perhaps a hundred million times smaller than the mass of the next-lightest particle in nature, the electron. Since the Standard Model does not accommodate neutrinos' masses, this was the first definitive observation that some new physics, beyond the Standard Model and beyond the Higgs, must be operating in nature.

Soon after this, observations of higher-energy neutrinos that regularly bombard Earth as high-energy cosmic-ray protons hit the atmosphere and produce a downward shower of particles, including neutrinos, demonstrated that yet a second neutrino has mass. This mass is somewhat larger, but still far smaller than the mass of the electron. For these results team leaders at SNO and Kamiokande were awarded the 2015 Nobel Prize in Physics—a week before I wrote the first draft of these words. To date these tantalizing hints of new physics are not explained by current theories.

The absence of proton decay, while disappointing, turned out to be not totally unexpected. Since Grand Unification was first proposed, the physics landscape had shifted slightly. More precise measurements of the actual strengths of the three nongravitational interactions—combined with more sophisticated calculations of the change in the strength of these interactions with distance—demonstrated that if the particles of the Standard Model are the only ones existing in nature, the strength of the three forces will not unify at a single scale. In order for Grand Unification to take place, some new physics at energy scales beyond those that have been observed thus far must exist. The presence of new particles would not only change the rate at which the three known interactions change with scale so that they might unify at a single scale of energy, it would also tend to drive up the Grand Unification scale and thus suppress the rate of proton decay—leading to predicted lifetimes in excess of a million billion billion billion years.

As these developments were taking place, theorists were driven by new mathematical tools to explore a possible new type of symmetry in nature, which became known as supersymmetry. This fundamental symmetry is different from any previous known symmetry, in that it connects the two different types of particles in nature, fermions (particles with half-integer spins) and bosons (particles with integer spins). The upshot of this (many other books, including some by me, explore this idea in detail) is that if this symmetry exists in nature, then for every known particle in the Standard Model at least one corresponding new elementary particle must exist. For every known boson there must exist a new fermion. For every known fermion there must exist a new boson.

Since we haven't seen these particles, this symmetry cannot be manifest in the world at the level we experience it, and it must be broken, meaning the new particles will all get masses that could be heavy enough so that they haven't been seen in any accelerator constructed thus far.

What could be so attractive about a symmetry that suddenly doubles all the particles in nature without any evidence of any of the new particles? In large part the seduction lay in the very fact of Grand Unification. Because if a Grand Unified Theory exists at a mass scale of fifteen to sixteen orders of magnitude higher energy than the rest mass of the proton, this is also about thirteen orders of magnitude higher than the scale of electroweak symmetry breaking. The big question is why and how such a huge difference in scales can exist for the fundamental laws of nature. In particular, if the Standard Model Higgs is the true last remnant of the Standard Model, then the question arises, Why is the energy scale of Higgs symmetry breaking thirteen orders of magnitude smaller-scale than the scale of symmetry breaking associated with whatever new field must be introduced to break the GUT symmetry into its separate component forces?

The problem is a little more severe than it appears. Scalar particles such as the Higgs have several new quantum mechanical properties that are unlike those of fermions or spin 1 particles such as gauge particles. When one considers the effects of virtual particles, including particles of arbitrarily large mass, such as the gauge particles of a presumed Grand Unified Theory, these tend to drive up the mass and symmetry-breaking scale of the Higgs so that it essentially becomes close to, or identical to, the heavy GUT scale. This generates a problem that has become known as the naturalness problem. It is technically unnatural to have a huge hierarchy between the scale at which the electroweak symmetry is broken by the Higgs particle and the scale at which the GUT symmetry is broken by whatever new heavy scalar field breaks that symmetry.

The brilliant mathematical physicist Edward Witten argued in an influential paper in 1981 that supersymmetry had a special property. It could tame the effect that virtual particles of arbitrarily high mass and energy have on the properties of the world at the scales we can currently probe. Because virtual fermions and virtual bosons of the same mass produce quantum corrections that are identical except for a sign, if every boson is accompanied by a fermion of equal mass, then the quantum effects of the virtual particles will cancel out. This means that the effects of virtual particles of arbitrarily high mass and energy on the physical properties of the universe on scales we can measure would now be completely removed.

If, however, supersymmetry is itself broken, then the quantum corrections will not quite cancel out. Instead they would yield contributions to masses that are the same order as the supersymmetry-breaking scale. If it was comparable to the scale of the electroweak symmetry breaking, then it would explain why the Higgs mass scale is what it is.

And it also means we should expect to begin to observe a lot of new particles—the supersymmetric partners of ordinary matter—at the scale currently being probed at the LHC.

This would solve the naturalness problem because it would protect the Higgs boson masses from possible quantum corrections that could drive them up to be as large as the energy scale associated with Grand Unification. Supersymmetry could allow a "natural" large hierarchy in energy (and mass) separating the electroweak scale from the Grand Unified scale.

That supersymmetry could in principle solve the hierarchy problem, as it has become known, greatly increased its stock with physicists. It caused theorists to begin to explore realistic models that incorporated supersymmetry breaking and to explore the other physical consequences of this idea. When they did so, the stock price of supersymmetry went through the roof. For if one included the possibility of spontaneously broken supersymmetry into calculations of how the three nongravitational forces change with distance, then suddenly the strength of the three forces would naturally converge at a single, very small-distance scale. Grand Unification became viable again!

Models in which supersymmetry is broken have another attractive feature. It was pointed out, well before the top quark was discovered, that if the top quark was heavy, then through its interactions with other supersymmetric partners, it could produce quantum corrections to the Higgs particle properties that would cause the Higgs field to condense at its currently measured energy scale if Grand Unification occurred at a much higher, superheavy scale. In short, the energy scale of electroweak symmetry breaking could be generated naturally within a theory in which Grand Unification occurs at a much higher energy scale. When the top quark was discovered and indeed was heavy, this added to the attractiveness of the possibility that supersymmetry breaking might be responsible for the observed energy scale of the weak interaction.

All of this comes at a cost, however. For the theory to work, there must be two Higgs bosons, not just one. Moreover, one would expect to begin to see the new supersymmetric particles if one built an accelerator such as the LHC, which could probe for new physics near the electroweak scale. Finally, in what looked for a while like a rather damning constraint, the lightest Higgs in the theory could not be too heavy or the mechanism wouldn't work.

As searches for the Higgs continued without yielding any results, accelerators began to push closer and closer to the theoretical upper limit on the mass of the lightest Higgs boson in supersymmetric theories. The value was something like 135 times the mass of the proton, with details to some extent depending on the model. If the Higgs could have been ruled out up to that scale, it would have suggested all the hype about supersymmetry was just that.

Well, things turned out differently. The Higgs that was observed at the LHC has a mass about 125 times the mass of the proton. Perhaps a grand synthesis was within reach.

The answer at present is . . . not so clear. The signatures of new supersymmetric partners of ordinary particles should be so striking at the LHC, if they exist, that many of us thought that the LHC had a much greater chance of discovering supersymmetry than it did of discovering the Higgs. It didn't turn out that way. Following three years of LHC runs, there are no signs whatsoever. The situation is already beginning to look uncomfortable. The lower limits that can now be placed on the masses of supersymmetric partners of ordinary matter are getting higher. If they get too high, then the supersymmetry-breaking scale would no longer be close to the electroweak scale, and many of the attractive features of supersymmetry breaking for resolving the hierarchy problem would go away.

But the situation is not yet hopeless, and the LHC has been turned on again, this time at higher energy. It could be that, in the year between the time I write these words and the book going into its tenth printing, supersymmetric particles will be discovered.

If they are, this will have another important consequence. One of the bigger mysteries in cosmology is the nature of the dark matter that appears to dominate the mass of all galaxies we can see. As I have briefly alluded to earlier, there is so much of it that it cannot be made of the same particles as normal matter. If it were, for example, the predictions of the abundance of light elements such as helium produced in the Big Bang would no longer agree with observation. Thus physicists are reasonably certain that the dark matter is made of a new type of elementary particle. But what type?

Well, the lightest supersymmetric partner of ordinary matter is, in most models, absolutely stable and has many of the properties of neutrinos. It would be weakly interacting and electrically neutral, so that it wouldn't absorb or emit light. Moreover, calculations that I and others performed more than thirty years ago showed that the remnant abundance today of the lightest supersymmetric particle left over after the Big Bang would naturally be in the range so that it could be the dark matter dominating the mass of galaxies.

In that case our galaxy would have a halo of dark matter particles whizzing throughout it, including through the room in which you are reading this. As a number of us also realized some time ago, this means that if one designs sensitive detectors and puts them underground, not unlike, at least in spirit, the neutrino detectors that already exist underground, one might directly detect these dark matter particles. Around the world a half dozen beautiful experiments are now going on to do just that. So far nothing has been seen, however.

So, we are in potentially the best of times *or* the worst of times. A race is going on between the detectors at the LHC and the underground direct dark matter detectors to see who might discover the nature of dark matter first. If either group reports a detection, it will herald the opening up of a whole new world of discovery, leading potentially to an understanding of Grand Unification itself. And if no discovery is made in the coming years, we might rule out the notion of a simple supersymmetric origin of dark matter—and in turn rule out the whole notion of supersymmetry as a solution of the hierarchy problem. In that case we would have to go back to the drawing board, except if we don't see any new signals at the LHC, we will have little guidance about which direction to head in order to derive a model of nature that might actually be correct.

Things got more interesting when the LHC reported a tantalizing possible signal due to a new particle about six times heavier than the Higgs particle. This particle did not have the characteristics one would expect for any supersymmetric partner of ordinary matter. In general the most exciting spurious hints of signals go away when more data are amassed, and about six months after this signal first appeared, after more data were amassed, it disappeared. If it had not, it could have changed everything about the way we think about Grand Unified Theories and electroweak symmetry, suggesting instead a new fundamental force and a new set of particles that feel this force. But while it generated many hopeful theoretical papers, nature seems to have chosen otherwise.

The absence of clear experimental direction or confirmation of supersymmetry has thus far not bothered one group of theoretical physicists. The beautiful mathematical aspects of supersymmetry encouraged, in 1984, the resurrection of an idea that had been dormant since the 1960s when Nambu and others tried to understand the strong force as if it were a theory of quarks connected by stringlike excitations. When supersymmetry was incorporated in a quantum theory of strings, to create what became known as superstring theory, some amazingly beautiful mathematical results began to emerge, including the possibility of unifying not just the three nongravitational forces, but all four known forces in nature into a single consistent quantum field theory.

However, the theory requires a host of new space-time dimensions to exist, none of which has been, as yet, observed. Also, the theory makes no other predictions that are yet testable with currently conceived experiments. And the theory has recently gotten a lot more complicated so that it now seems that strings themselves are probably not even the central dynamical variables in the theory.

None of this dampened the enthusiasm of a hard core of dedicated and highly talented physicists who have continued to work on super-string theory, now called M-theory, over the thirty years since its heyday in the mid-1980s. Great successes are periodically claimed, but so far M-theory lacks the key element that makes the Standard Model such a triumph of the scientific enterprise: the ability to make contact with the world we can measure, resolve otherwise inexplicable puzzles, and provide fundamental explanations of how our world has arisen as it has. This doesn't mean M-theory isn't right, but at this point it is mostly speculation, although well-meaning and well-motivated speculation.

Here is not the place to review the history, challenges, and successes of string theory. I have done that elsewhere, as have a number of my col-leagues. It is worth remembering that if the lessons of history are any guide, most forefront physical ideas are wrong. If they weren't, anyone could do theoretical physics. It took several centuries or, if one counts back to the science of the Greeks, several millennia of hits and misses to come up with the Standard Model.

So this is where we are. Are great new experimental insights just around the corner that may validate, or invalidate, some of the grander speculations of theoretical physicists? Or are we on the verge of a desert where nature will give us no hint of what direction to search in to probe deeper into the underlying nature of the cosmos? We'll find out, and we will have to live with the new reality either way.

No matter what curveballs nature may throw at us, the recent dis-covery of the Higgs, the latest and one of the greatest experimental and theoretical achievements of the remarkable Standard Model of particle physics, has beautifully capped more than two millennia of intellectual effort by brave and determined philosophers, mathematicians, and sci-entists to uncover the hidden tapestry that underlies our existence.

It also suggests that the beautiful universe in which we find ourselves may not only resemble, at least metaphorically, an ice crystal on a win-dowpane, it may be almost as ephemeral.

Chapter 23

FROM A BEER PARTY TO THE END OF TIME

For the fashion of this world passeth away.

—1 CORINTHIANS 7:31

My own research focus for much of my career has been the emerging field of cosmology called particle astrophysics. Following the flood of theoretical developments of the 1960s and 1970s, it was difficult for terrestrial experiments, limited as they are by our abilities to build complex machines such as particle accelerators, to keep up. As a result, a number of us turned to the universe for guidance. Since the Big Bang implies that the early universe was hot and dense, conditions existed then that we might never achieve in laboratories on Earth. But if we are clever, we can look for remnant signatures of those early times out in the cosmos, and we may be able test our ideas about even the most esoteric aspects of fundamental physics.

My previous book, *A Universe from Nothing*, described the revolutions in our understanding of the evolution of the universe on large scales, and over long times. Not only have our explorations revealed the existence of dark matter, which, as I have described, is likely composed of new elementary particles not yet observed in accelerators—although we may be on the cusp of doing so—but far more exotic still, we have discovered that the dominant energy of the universe resides in empty space—and we currently have no idea how it arises.

Our observations have now taken us back to the neonatal universe. We have observed the fine details of radiation, called the cosmic microwave background, which emanates from a time when the universe was merely three hundred thousand years old. Our telescopes take us back to the earliest galaxies, which formed perhaps a billion years after the Big Bang, and have allowed us to map huge cosmic structures containing thousands of galaxies and spanning hundreds of millions of light-years across, sprinkled amid the hundred billion or so galaxies in the visible universe.

To explain these features, theorists rely on an idea that arose due to the development of Grand Unified theories. In 1981, Alan Guth realized that the symmetry-breaking transition that might occur at the GUT scale early in the universe might not be identical to the transition that breaks the symmetry between the weak interaction and electromagnetism. In the GUT case, the Higgs-like field that condenses in space to break the GUT symmetry between the strong force and the electroweak force might momentarily get stuck in a metastable high-energy state before relaxing to its final configuration. While it was in this "false vacuum" configuration, the field would store energy that would be released when the field ultimately relaxed to its preferred lowest-energy configuration.

The situation would not be unlike what may have happened to you if you have ever planned a big party and then forgotten to put the beer in the fridge in time. You then put the beer in the freezer and forget about it during the party. The next day you discover the beer, open a bottle, and *wham!* The beer in the bottle suddenly freezes and expands, shattering the glass, and producing quite a mess. Before the top is taken off, the beer is under high pressure, and the beer at this pressure and temperature is liquid. However, once you open the top and release the pressure, the beer suddenly freezes. During the transition, energy is released as the beer relaxes to its new state—enough energy to cause the expanding ice to break the bottle.

Now imagine a similar situation when you are in a cold climate. On a brisk and rainy winter day, the temperature may quickly drop below freezing, causing the rain to change to snow. Puddles of water on the street may not freeze right away, especially if the tires of passing cars are continually agitating them. Later in the day, when the traffic dies down, the water may suddenly freeze, causing dangerous black ice on the road. Due to the previous agitation by cars and the quick fall in temperature, the water got stuck in a "metastable phase," namely as a liquid. Eventually, however, a phase transition takes place, and the black ice forms. Because at these low temperatures the preferred, lowest-energy state of water is its solid form, when the liquid freezes, it releases the excess energy it stored in its metastable liquid state.

Guth wondered what would have happened in the early universe if such a behavior occurred during a Grand Unified Theory transition—if whatever scalar field that acts like the Higgs field for that transition remains in its original (symmetry-preserving) ground state for a brief time, even as the universe cools past the point where the new (symmetry-breaking) ground state condensate becomes preferred. Guth realized that this type of energy, stored through space by this field before the transition completes, would be gravitationally repulsive. As a result, it would cause the universe to expand—potentially by a huge factor, maybe twenty-five orders of magnitude or more in scale—in a microscopically short time.

He next discovered that this period of rapid expansion, which he dubbed inflation, could resolve a number of existing paradoxes associated with the Big Bang picture, including why the universe is so uniform on large scales and why three-dimensional space on large scales appears so close to being geometrically flat. Both of these seem inexplicable without inflation. The first problem is solved because, during the rapid expansion, any initial inhomogeneities get smoothed out, just as a wrinkled balloon gets smoothed out when it gets blown up. Pushing the balloon analogy further, the surface of a balloon that is blown up to be very large, say, the size of Earth, could look very flat, just as Kansas does. While this provides two-dimensional intuition, the same phenomenon would apply to the three-dimensional curvature of space itself. After inflation, space would appear to be flat—namely it would be like the universe most of us had assumed we live in already, where parallel lines never intersect and the x, y, and z axes point the same direction everywhere in the universe.

After inflation ends, the energy stored in the false vacuum state throughout space would be released, producing particles and reheating the universe to a high temperature, setting up a natural and realistic initial condition for the subsequent standard hot Big Bang expansion.

Even better, a year after Guth proposed his picture, a number of groups performed calculations of what would happen to particles and fields as the universe rapidly expanded during inflation. They discovered that small inhomogeneities resulting from quantum effects at early times would then be "frozen in" during inflation. After inflation ended, these small inhomogeneities could grow to produce galaxies, stars, planets, etc., and would also leave an imprint in the cosmic microwave background (CMB) radiation that resembles precisely the pattern that has since been measured. However, it is also possible, by using different inflationary models, to get different predictions for the CMB anisotropies (inflation is, at this point, more of a model than a theory, and since no unique Grand Unified Theory transition is determined by experiment, many different variants might work).

Another exciting and more unambiguous prediction from inflation exists. During the period of rapid expansion, ripples in space, called gravitational waves, would be produced. These ripples would produce another characteristic signature in the CMB that might be sought out. In 2014, the BICEP experiment claimed to detect a signal that was identical to what was predicted. This caused incredible excitement in both theoretical and observational communities. Along with Frank Wilczek, I wrote a paper that not only pointed out that such an observation would indicate a symmetry-breaking scale that corresponded nicely to the Grand Unified Theory symmetry-breaking scale in models with supersymmetry, but also that the observation would demonstrate unambiguously that gravity had to be a quantum theory on small scales—so that a search for a quantum theory of gravity was not misplaced.

Unfortunately, however, the BICEP announcement proved to be premature. Other backgrounds in our galaxy could have produced a similar signal, and as of this writing the situation still seems murky, with no unambiguous confirmation of inflation, or quantum gravity.

Most recently, between completion of the first draft of this book and completion of the final draft, the first definitive direct discovery of gravitational waves was made by an amazing set of detectors, called the Laser Interferometer Gravitational-Wave Observatory (LIGO), located in Hanford, Washington, and Livingston, Louisiana. LIGO is a spectacular and ambitious machine. To detect gravitational waves emitted by colliding black holes in distant galaxies, the experimenters had to be able to detect an (oscillating) difference in length between two four-kilometer-long perpendicular arms of the detectors equivalent to one one-thousandth of the size of a proton—like measuring the distance between Earth and the nearest star other than our Sun, Alpha Centauri, to an accuracy of the width of a human hair!

As amazing as the LIGO discovery of gravitational waves is, the waves it detected are from a distant astrophysical collision, not from the earliest moments of the Big Bang. But the success of LIGO will herald the building of new detectors, so that gravitational-wave astronomy will likely become the astronomy of the twenty-first century.

If the successors to LIGO, or BICEP, in this or the next century are able to measure directly the signature of gravitational waves from inflation, it will give us a direct window on the physics of the universe when it was less than a billionth of a billionth of a billionth of a billionth of a second old. It will allow us to directly test our ideas of inflation, and even Grand Unification, and perhaps even shed light on the possible existence of other universes—turning what is now metaphysics into physics.

For the moment, however, inflation is merely a well-motivated proposal that seems to naturally resolve most of the major puzzles in cosmology. But while inflation remains the only first-principles theoretical-candidate explanation for the major observational features of our universe, it relies on the existence of a new and completely ad hoc scalar field—invented solely to help produce inflation and fine-tuned to initiate it as the early universe first began to cool down after the Big Bang.

Before the discovery of the Higgs particle, this speculation was plausible at best. With no example of any fundamental scalar field yet known, the assumption that Grand Unified symmetry-breaking might result from *yet another* simple Higgs-like mechanism was an extrapolation that rested on an insecure footing. As I have described, the breaking of electroweak symmetry was clear with the discovery of W and Z particles. But the simple Higgs field could have been a fairy-tale placeholder for some far more complicated, and perhaps far more interesting, underlying mechanism.

Things have now changed. The Higgs exists, and so too apparently a background scalar field permeating all space in the universe today, giving mass to particles and producing the characteristics of a universe we can inhabit. If a Grand Unified Theory really exists combining all three forces into one at close to the beginning of time, some symmetry breaking must have then occurred so that the three known nongravitational forces would only begin to diverge in character afterward. The Higgs demonstrates that symmetry breaking in the laws of nature can occur as the result of a scalar field condensate throughout space. Depending upon the details, inflation thus becomes a far more natural and potentially generic possibility. As my colleague Michael Turner put it jokingly some time ago, aping then Federal Reserve Board chair Alan Greenspan, "Periods of inflation are inevitable!"

That statement may have been more prescient than anyone imagined at the time. In 1998 it was discovered that our universe is now undergoing a new version of inflation, validating some previous and rather heretical predictions by a few of us. As I mentioned earlier, this implies that the dominant energy of the universe now appears to reside in empty space—which is the most plausible explanation of why the observed expansion of the universe is speeding up. The Nobel Prize was awarded to Brian Schmidt, Adam Riess, and Saul Perlmutter for the discovery of this remarkable and largely unexpected phenomenon. Naturally the questions arise, What could be causing this current accelerated expansion, and What is the source of this new kind of energy?

Two possibilities present themselves. First, it could be a fundamental property of empty space, a possibility actually presaged by Albert Einstein shortly after he developed the General Theory of Relativity, which he realized could accommodate something he called a "cosmological constant," but which we now realize could simply represent a nonzero ground-state energy of the universe that will exist indefinitely into the future.

Or second, it could be energy stored in yet another invisible background scalar field in the universe. If this is the case, then the next obvious question is, will this energy be released in yet another, future inflationary-like phase transition as the universe continues to cool down?

At this time the answer is up for grabs. While the inferred energy density of empty space is today greater than the energy density of everything else we see in the universe, in absolute terms, on the scale of the energies associated with the masses of all elementary particles we know of, it is minuscule in the extreme. No one has any sensible first-principles explanation using known particle physics mechanisms for how the ground-state energy of the universe could be nonzero—resulting in Einstein's cosmological constant—and yet so small as to allow the kind of gentle acceleration we are now experiencing. (One plausible explanation does exist—first due to Steve Weinberg—though it is speculative and relies on speculative ideas about possible physics well beyond the realm of anything we currently understand. If there are many universes, and the energy density in empty space, assuming it is a cosmological constant, is not fixed by fundamental physics constraints, but instead randomly varies from universe to universe, then only in those universes in which the energy in empty space is not much bigger than the value we measure would galaxies be able to form, and then would stars be able to form, and only then planets, and only then astronomers . . .)

Meanwhile, no one has a sensible model for a new phase transition predicted to occur in particle physics for a new scalar field that would store such a small amount of energy in space today. By sensible, I mean a model that anyone other than those who propose it finds plausible.

Nevertheless, the universe is the way it is, and the fact that current fundamental theory does not make a first-principles prediction that explains something as fundamental as the energy of empty space implies nothing mystical. As I have said, lack of understanding is *not* evidence for God. It is merely evidence of a lack of understanding.

Given that we do not know the source of the inferred energy in empty space, we are free to hope for the best, and in this case perhaps that means hoping that the cosmological constant explanation is correct rather than its being due to some as yet undiscovered scalar field that may one day relax into a new state, releasing the energy currently stored in space.

Recall that because of the coupling of the Higgs field to the rest of the matter in the universe, when the field condensed into its electroweak symmetry-breaking state, the properties of matter and the forces that govern the interactions of matter changed dramatically.

Now, if some similar phase transition involving some new scalar field in space is yet to occur in nature, then the stability of matter as we know it could disappear. Galaxies, stars, planets, people, politicians, and everything we now see could literally disappear. The only good news (other than the disappearance of politicians) is that the transition—assuming it begins with some small seed in one location of our universe (in the same way that small dust grains may help seed the formation of the ice crystals on our frozen windowpane, or of snowflakes as they fall to the ground)—will then spread throughout space at the speed of light. We won't know what hit us until after it has, and after it has, we won't be around to know.

The curious reader may have noticed that all of these discussions relate to new possible scalar fields in nature. What about the Standard Model Higgs field? Could it play a role in all of these current cosmic shenanigans? Could the Higgs field store energy and be responsible for inflation either in the early universe or now? Could the Higgs field not be in its final ground state, and will there be another transition that will once again change the configuration of the electroweak force, and the masses of particles in the Standard Model?

Good questions. And the answers to all of them are the same: we don't know.

That has not stopped a number of theorists from speculating about this possibility. My favorite example—not because it is better than any of the others, but because it's a speculation I made with a colleague, James Dent, shortly after the Higgs was discovered—is that perhaps the Higgs does play a role in the observed cosmic expansion. As a number of authors have recognized, the existence of one background field condensate and the particles it comprises can provide a unique window, or "portal," that may yield otherwise unexpected sensitivity to the existence of other Higgs-like fields in nature, no matter how weakly their direct couplings to the particles we observe in the Standard Model may be.

If the Higgs and other Higgs-like particles exist, perhaps at the Grand Unified Theory scale, the physical Higgs, the particle that was discovered at CERN, may be a slight admixture between the weak interaction Higgs, and another Higgs-like particle. (In this we are guided by the physics of neutrinos, where similar phenomena play a vital role in understanding the behavior of neutrinos measured on Earth coming from the nuclear reactions in the Sun, for example.) It is then possible, at least, to argue that when the weak interaction Higgs field condenses in empty space, this could stimulate the condensation of another Higgs-like field with properties that would allow it to store just the right energy to explain the observed inflation of the universe today. The mathematics required to make this happen is pretty contrived—the model is ugly. But who knows? Maybe it is ugly because we haven't found the correct framework in which to embed it.

However, one attractive feature of this scenario makes it a little less self-serving to mention it. In this picture, the energy carried by the second field, which would drive the current measured accelerated expansion of the universe today, will likely ultimately be released in a new phase transition to the true ground state of the universe. Unlike many other possibilities for future possible phase transitions in our universe, because the new field can be weakly coupled to all observed particles, this transition will not induce a change in the observed properties of any of the known particles in nature by an amount that would be noticeable. The upshot is that if this model is right, the universe as we know it may survive.

Yet celebration may be premature. Independent of such speculations, the discovery of the Higgs particle has raised the specter of a much less optimistic possibility. While a future in which the observed acceleration of the universe goes on forever is a miserable future for life and for the ability to continue to probe the universe—because eventually all galaxies we can now observe will recede from us faster than light, ultimately disappearing from our horizon, leaving the universe cold, dark, and largely empty—the future that may result because of a Higgs field with a mass 125 times the mass of the proton could be far worse.

For a Higgs mass coinciding with the allowed range of the observed

Higgs, assuming for the moment that the Standard Model is not supplemented by a lot of new stuff at higher energy, calculations suggest that the existing Higgs field condensate is teetering on the edge of instability—it could change from its current value to a vastly different value associated with a lower-energy state.

If such a transition occurs, normal matter as we know it changes its form, and galaxies, stars, planets, and people most likely disappear, like the ice crystal on a warm sunny morning.

For those who enjoy horror stories, another, even more gruesome possibility has been suggested. An instability might exist that would cause the Higgs field to continue to grow in magnitude indefinitely. As a result of such growth, the energy stored by the evolving Higgs field could become negative. This could cause the entire universe to collapse once again in a cataclysmic reversal of the Big Bang—a Big Crunch. Happily the data disfavor such a possibility, as poetic as it might seem.

In the scenario in which everything we now see disappears as the Higgs makes a sudden transition to a new ground state, I want to stress that the Higgs mass, as now measured, favors stability but has sufficient uncertainty in its value to fall on either side of this line—either producing the apparently stable vacuum that we are now flourishing in, or favoring such a transition. Moreover, this scenario is based on calculations within the Standard Model alone. Any new physics that might be discovered at the LHC or beyond could change the picture entirely, stabilizing what could otherwise be an unstable Higgs field. Since we are reasonably certain there is new physics to be discovered, there is no cause for despair at present.

If that isn't consolation enough, for those who still fear that the ultimate future of the universe might be the more miserable one I have just described, the same calculations that suggest this may happen also suggest that our current metastable configuration of reality would persist for not merely billions of years into the future, but billions of billions of billions of years.

Concerns about the future notwithstanding, now is an appropriate time to once again emphasize that the universe doesn't give a damn what we would like or whether we survive. Its dynamics continue independently of whether we exist or not. For this reason I am strangely attracted to the doomsday scenario I have just described. In this case, the remarkable accident that is responsible for our existence—the condensation of a field that allows the current stability of matter, atoms, and life itself—is seen as a short-term bit of good luck.

The imaginary scientists living on the spine of an ice crystal on the windowpane that I described earlier would first discover that one direction in their universe was particularly special (which would no doubt be celebrated by the theologians in such a society as an example of God's love). Digging deeper, they might discover that this special circumstance is just an accident and that other ice crystals can exist in which other directions are favored.

And so, we too have discovered that our universe, with its forces and particles and amazing Standard Model that results in the remarkable good fortune of an expanding universe with stars and planets and life that can evolve a consciousness, is also a simple accident made possible because the Higgs field condensed in just the way it did as the universe evolved early on.

And even as the imaginary scientists on the hypothetical ice crystal might celebrate their discoveries as we are wont to do, they might also be unaware that the Sun is about to rise and that soon their ice crystal will melt, and all traces of their brief existence will disappear. Would this have made the thrill of their brief existence less enthralling? Certainly not. If our future is similarly fleeting, we can at least enjoy the wild ride we have taken and relish every aspect of the greatest story ever told ... so far.

COSMIC HUMILITY

For dust thou art, and unto dust shalt thou return.

—GENESIS 3:19

"These are the tears of things, and the stuff of our mortality cuts us to the heart."

So said Virgil as he penned the first great epic story of the classical era. They are the words I chose to use as the epigraph of this book because the story I wanted to tell not only contains every bit as much drama, human tragedy, and exaltation, but it is ultimately motivated by a similar purpose.

Why do we do science? Surely it is in part so that we can have greater control of our environment. By understanding the universe better we can predict the future with greater accuracy, and we can build devices that might change the future—hopefully for the better.

But ultimately I believe we are driven to do science because of a primal urge we have to better understand our origins, our mortality, and ultimately ourselves. We are hardwired to survive by solving puzzles, and that evolutionary advantage has, over time, allowed us the luxury of wanting to solve puzzles of all sorts—even those less pressing than how to find food or to escape from a lion. What puzzle is more seductive than the puzzle of our universe?

Humanity didn't have a choice in its evolution. We find ourselves alive on a planet that is 4.5 billion years old in a galaxy that is 12 billion years old, in a 13.8-billion-year-old universe with at least a hundred billion galaxies that is expanding ever faster into a future we cannot yet predict.

So what do we do with this information? Is there special significance here for understanding our human story? In the midst of this cosmic grandeur and tragedy, how can we reconcile our own existence?

For most people, the central questions of existence ultimately come down to transcendental ones: Why is there a universe at all? Why are we here?

Whatever presumptions one might bring to the question "Why?," if we understand the "how" better, "why" will come into sharper focus. I wrote my last book to address what science has to say about the first of the above questions. The story I have related here provides what I think is the best answer to the second.

Faced with the mystery of our existence, we have two choices. We can assume we have special significance and that somehow the universe was made for us. For many, this is the most comfortable choice. It was the choice made by early human tribes, who anthropomorphized nature because it provided them some hope of understanding what otherwise seemed to be a hostile world often centered on suffering and death. It is the choice made by almost all the world's religions, each of which has its own claimed solution to the quandary of existence.

This choice of which tale to embrace has led to one culture's sacred book, the New Testament, which has sometimes been called "the greatest story ever told"—the story of that civilization's putative discovery of its own divinity. Yet when I witness wars and killing based on which prayers we are supposed to recite, which persons we are supposed to marry, or which prophet is the appropriate one to follow, I cannot help but be reminded, once again, of Gulliver, who discovered societies warring over which way God had intended man to break an egg.

The second choice when addressing these transcendental mysteries is to make no assumption in advance about the answer. Which leads to another story. One that I think is more humble. In this story we evolve in a universe whose laws exist independently of our own being. In this story we check the details to see if they might be wrong. In this story we are going to be surprised at every turn.

The story I have written here describes a human drama as much as a universal one. It describes the boldest intellectual quest humans have ever undertaken. It even has scriptural allegories, for those who prefer them. We wandered in the desert for forty years after the development of the Standard Model before we discovered the Promised Land. The truth, or at least as much of the truth as we now know, was revealed to us in what for most people seems to be incomprehensible scribbles: the mathematics of gauge theories. These have not been delivered to us on golden tablets by an angel, but rather by much more practical means: on pieces of paper in laboratory notebooks filled through the hard work of a legion of individuals who knew that their claims could be tested by whether they correctly modeled the real world, the world of observation and experiment. But as significant as the manner by which we got here is that we have gotten this far.

At this point in the story, what can we conclude about why we are here? The answer seems all the more remarkable because it reveals explicitly just how deeply the universe of our experience is a shadow of reality.

I also began this book with a quote from the naturalist J. A. Baker, from *The Peregrine*: "The hardest thing of all to see is what is really there." I did so because the story I have told is the most profound example of this wise observation that I know of.

I next described Plato's Allegory of the Cave because I know of no better or more lyrical representation of the actual history of science. The triumph of human existence has been to escape the chains that our limited senses have imposed upon us. To intuit that beneath the world of our experience lies a reality that is often far stranger. It is a reality whose mathematical beauty may be unimpeachable, but a reality in which our existence becomes—more than we might ever have imagined in advance—a mere afterthought.

If we now ask why things are the way they are, the best answer we can suggest is that it is the result of an accident in the history of the universe in which a field froze in empty space in a certain way. When we ponder what significance that might have, we might equally ponder what is the significance of that specific ice crystal seen in the early-morning frost on a windowpane. The rules that allowed us to come into being seem no more worth fighting and dying for than it would seem to be to fight and die to resolve whether "up" in the ice-crystal universe is better than "down," or whether it is better to crack an egg from the top or the bottom.

Our primitive ancestors survived in large part because they recognized that nature could be hostile and violent, even as it was remarkable. The progress of science has made it clear just how violent and hostile the universe can be for life. But recognizing this does not make the universe less amazing. Such a universe has ample room for awe, wonder, and excitement. If anything, recognition of these facts gives us greater reason to celebrate our origins, and our survival.

To argue that, in a universe in which there seems to be no purpose, our existence is itself without meaning or value is unparalleled solipsism, as it suggests that without us the universe is worthless. The greatest gift that science can give us is to allow us to overcome our need to be the center of existence even as we learn to appreciate the wonder of the accident we are privileged to witness.

Light played a major role in our story, as it did in Plato's allegory. Our changing perception of light led us to a changing understanding of the essence of space and time. Ultimately that changing perception made it clear that even this messenger of reality that is so essential to us and our existence is itself merely a fortunate consequence of a cosmic accident. An accident that may someday be rectified.

It is appropriate here to recognize that the line in the *Aeneid* that follows the epigraph with which this book began was the hopeful cry "Release your fear." A future that might bring about our end does not negate the majesty of the journey we are still taking.

The story I have told is not the whole story. There is likely to be far more that we don't understand than what we now do. In the search for meaning, our understanding of reality will surely change as the story continues to unfold. I am often told that science can never do some things. Well, how do we know until we try?

As fate would have it, I am writing these final words while sitting at the desk at which my late friend and coconspirator in the battle against myth and superstition Christopher Hitchens wrote his masterpiece, *God Is Not Great.* It is hard not to feel his presence channeling these words, even as I know he would be the first to remind me that such feelings arise from inside my head, and not from anything more cosmically significant. Yet the title of his book emphasizes that human stories, which he loved so dearly and described so brilliantly, pale in comparison to the story that nature has driven us to discover. And so the human stories about God also pale in comparison to the real "greatest story ever told."

This story ultimately does not give the past special significance. We can reflect upon and even celebrate the road we have taken, but the greatest liberation, and the greatest solace that science provides, come from perhaps its greatest lesson: that the best parts of the story can yet be written.

Surely this possibility makes the cosmic drama of our existence worthwhile.

ACKNOWLEDGMENTS

This book is written in part as a tribute to all of those who have helped bring our understanding of the universe to the place it is today. Because I wanted to properly and appropriately represent the science, and the history, to help me check both I turned to a number of my colleagues after I finished the first version of this book. I received comments and useful suggestions and corrections in response, and I want to thank both Sheldon Glashow and Wally Gilbert for their suggestions, as well as Richard Dawkins, and I am particularly indebted to one of the colleagues I admire most for his contributions as a scientist and his scientific integrity, who would rather remain anonymous, for his careful reading of the manuscript, and the numerous corrections he proposed. Beyond the science, I turned to a friend and one of the writers I admire most, who is also a wonderful student of science, for his thoughts on the manuscript. Cormac McCarthy, who amazingly volunteered to copyedit the paperback version of my earlier book *Quantum Man*, again went through every single page of the manuscript he received, with comments and suggestions to, in his words, "make the book perfect." I cannot presume that it now is, but I can say that it is much better thanks to his kindness, wisdom, and talent.

This book would never have been written if determining a publisher hadn't been skillfully managed by my new agent and old friend John Brockman and his staff, and happily it worked out that my editor for this book was my editor for *A Universe from Nothing*, Leslie Meredith at Atria Books. Leslie is not only a kindred spirit, but was a wonderful foil off of which to bounce the ideas in this book. She helped force me to make various discussions of the science clearer, even when I thought they were already clear, and she encouraged me not to back off from my strong views on the need for scientists to speak out about scientific nonsense.

When I faced the arduous task of exploring a variety of significant revisions in the final draft, I knew that I could seek safety, support, and solitude in the home that my wonderful wife, Nancy, who has saved me and inspired me more times than I can count, has made for us, and that my stepdaughter, Santal, would quietly tolerate the sound of my typing in my study, right above her bedroom, late at night. My staff at the Origins Project, in particular my executive director and right-hand woman, Amelia Huggins, and my longtime executive assistant at Arizona State University, Jessica Strycker, pitched in to provide me the support and time I needed when I had to take time out from my day job to work on this book. And my Phoenix friends Thomas Houlon and Patty Barnes, who encouraged me on this book and others, have, over many breakfasts, given their feedback on a number of the presentations I developed as I was writing the book.

Finally, as I was approaching the last push, my friend Carol Blue, Christopher Hitchens's widow, and her father, Edwin Blue, offered me use of a guesthouse where Christopher had written many essays and books, including his wonderful book *God Is Not Great*. I cannot think of a more inspiring place to have finished, and I can only hope the final version carries with it even a small fraction of the eloquence that so characterized Christopher's writing.

INDEX

ABOUT THE AUTHOR

Lawrence M. Krauss is director of the Origins Project at Arizona State University and Foundation Professor in the School of Earth and Space Exploration and the Physics Department there. Krauss is an internationally known theoretical physicist with wide research interests, including the interface between elementary-particle physics and cosmology, where his studies include the early universe, the nature of dark matter, general relativity, and neutrino astrophysics. He has investigated questions ranging from the nature of exploding stars to issues of the origin of all mass in the universe. He has won numerous international awards for both his research and his efforts to improve the public understanding of science. Krauss is the only physicist to have received the top awards from all three US physics societies: the American Physical Society, the American Institute of Physics, and the American Association of Physics Teachers, and in 2012 he was awarded the National Science Board's prestigious Public Service Award for his many contributions to public education and the understanding of science around the world. Among his other honors are the 2013 Roma Award, from the city of Rome, and the 2015 Humanist of the Year Award from the American Humanist Association.

Krauss is the author of more than three hundred scientific publications, as well as numerous popular articles on science and current affairs. He is a commentator and essayist for periodicals such as the *New York Times* and the *New Yorker* and appears regularly on radio, on television, and on film. Krauss served as executive producer and subject of *The Unbelievers*, a documentary film that discusses science and reason with Richard Dawkins. He also appears in Werner Herzog's new films *Salt and Fire* and *Lo and Behold*. Krauss is the author of ten popular books, including the *New York Times* bestsellers *The Physics of Star Trek* (1995) and *A Universe from Nothing* (2012).

Krauss is a Fellow of the American Physical Society and the American Association for the Advancement of Science. He serves as the chair of the Board of Sponsors of the *Bulletin of the Atomic Scientists* and is on the Board of Directors of the Federation of American Scientists. He helped found ScienceDebate, which in 2008, 2012, and 2016 helped raise issues of science and sound public policy in the presidential elections in those years. Hailed by *Scientific American* as a rare scientific public intellectual, Krauss has dedicated his time, throughout his career, to issues of science and society and has helped spearhead national efforts to educate the public about science, ensure sound public policy, and defend science against attacks at a variety of levels.